M

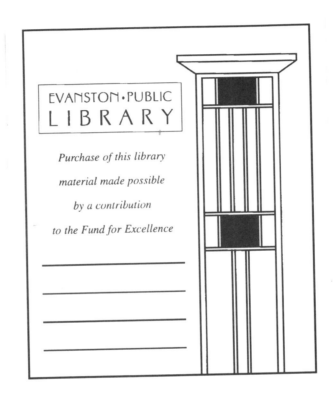

30 Bridges

30 Bridges Matthew Wells / Introduction by Hugh Pearman

Watson-Guptill Publications / New York

First published in 2002 in the United States by
Watson-Guptill Publications,
a division of BPI Communications, Inc.
770 Broadway, New York, NY 10003
www.watsonguptill.com

First published in 2002
by Laurence King Publishing Ltd
71 Great Russell Street
London WC1B 3BP

Library of Congress Control Number: 2001089840

ISBN 0-8230-5354-7

Designed by Frank Philippin
at Brighten the Corners

Printed in Singapore

1 2 3 4 5 6 7 8 9 / 10 09 08 07 06 05 04 03 02

Contents

Introduction Hugh Pearman

Towards the end of the twentieth century, deep into the era of post-war value engineering, bridges were rediscovered as vessels of metaphor. A bridge, it was realized, was more than a matter of cold calculation, a way of spanning a given gap to carry a given variable load in certain predictable conditions, using the most economic means. A bridge could be – had previously been – more. Of course the symbolic and political function of linkage had never been overlooked: for instance the great suspension bridges of the 1960s and 1970s had been audacious and beautiful enough to provoke wonder. But at the turn of the twenty-first century, bridges have come to be perceived in other ways. As a result, what had been largely the kingdom of the traditional civil engineer has now also become a playground for architects and a new breed of design engineers.

Much of the appeal of the bridge type as a design challenge derives from its chameleon nature. Is a bridge a building, or a room, or a road or pathway, or an observation deck, or a gateway, or a roof, or a monument? To what extent is it functionalist rather than expressive? If expressive, to what extent is that communicated through its load-carrying structure, and to what extent through applied forms? It is very likely that nobody ever sits down with such a checklist when they want a bridge. But they have certain desires, and those desires are communicated, for creative interpretation, to their designers.

At the bottom end of the range, a motorway overbridge, say, is required by the highway authorities. Or more likely, a clutch of dozens of such overbridges. At their most basic, such structures seem to be merely extruded pieces of road slab. At their best, they give a distinct character to a route: but these are the workhorse bridges, the equivalent of the hump-backed brick and stone canal access bridges of the eighteenth century. Low maintenance, good manners and a certain finesse is the most one can normally expect. In contrast, at the top of the range, a bridge is required that will serve a symbolic function far beyond its quotidian use. Such was the Erasmus Bridge of 1996 in Rotterdam by Ben van Berkel and others, intended as a catalyst for regeneration in the run-down docks area of the city, a memorably sculpted vertical marker in the horizontal townscape. As with Florida's famous 1987 Sunshine Skyway (engineers Figg and Muller) with its yellow-sheathed cables making sail shapes across the water, the visual role of the Erasmus Bridge was clear from the outset.

The case is different at the Charles River crossing (2001) in Boston. The bridge, by the Swiss engineer Christian Menn, is sited in an historic area where Paul Revere began his famous ride and the Battle of Bunker Hill took place. Here, the task that the bridge is asked to perform is to some extent contradictory. On the one hand, it was acknowledged that this sensitive site deserved a considered, even reticent, design. On the other, this was going to be the main route into Boston from the north. The phrase 'dramatic gateway to the city' was bandied about. Sixteen bridge types were considered, eventually whittled down to a choice of two cable-stay options: one with a single tall landmark tower in the style of Rotterdam, the other – Christian Menn's solution – with smaller twin towers. This latter option was selected as the best compromise between drama and deference. Much of the aesthetic success of the design comes down to a level of detail relatively rare in highway bridge construction. Openings were made in the wide main span deck, to avoid heavy shadow on the water beneath. Cladding panels give the bridge a smooth underbelly. There is an alternation of splayed and centred cables across the spans.

Such large-scale roadway landmarks stand in clear line of descent from the barbican towers of medieval cities. The fortified bridge at Cahors in central France (fig. 1), with its triple tall defence towers, looks to our modern eyes as if it is crying out for cables to be slung

from those towers to suspend the deck, as chains might suspend a drawbridge. The sculpting of today's support towers for suspension and cable-stay bridges satisfyingly fulfils the Cahors craving: vertical elements that not only exactly balance the horizontal, but which are also absolutely necessary to it. There is no better example of the moment of transition between the one type and the other than Tower Bridge in London (fig. 2) of 1890 by Sir Horace Jones and J. Wolfe Barry. This was a modern steel structure combining suspended decks, steam-driven rising bascules and daringly engineered twin high-level walkways for pedestrians – yet Barry's engineering was clad by Jones in a mock-medieval granite carapace recalling those ancient bastions. What would it have looked like without the stone cladding? To judge by the internal structure, utilitarian in the extreme. Certainly nowhere near as good as the open latticework of Othmar Ammann's later George Washington Bridge over the Hudson River, the towers of which were originally intended to be clad in granite-faced concrete. The Depression during of the inter-war years meant that the masonry skin was cancelled, to great visual effect.

This division of labour – engineers for the structure, architects for applied form, laid on like piecrust – goes back a long way. The Wearmouth Bridge of 1796 (fig. 3) – an early example of an iron road bridge, considered a wonder in its day – was promoted by the politician and entrepreneur Rowland Burdon, and designed by his friend the architect Sir John Soane with the engineer Thomas Wilson. Soane, the most technically adept classicist of his period, was as much the project manager as the designer: he brought in Wilson to refine existing iron-bridge ideas by Tom Paine and others, and built it. Most bridges are still essentially designed by engineers, and only then is the form aestheticised to some extent by architects. Such grand affairs as Foster's Grand Viaduc du Millau (fig. 4), being built not far from Cahors, falls into this category: the engineering parameters were set before architects were invited to compete for the design. But there was a real architectural competition, and the architectural input was considerable. More usual, however, is the case of the William Natcher cable-stay bridge (fig. 5; 2002) carrying Route 231 over the Ohio River between Kentucky and Indiana. This, one of the larger cable-stay bridges in the USA, was designed by Vijay Chandra and others of the engineering firm Parsons Brinckerhoff. Chandra's description of the genesis of the project, given at a conference in 1999, is instructive. The shape of the towers, he remarked, was selected on the basis of economy, functionality, constructibility, inspectability, ease of maintenance, torsional stability, cable connectivity and so forth. Once all that was agreed, the bridge was handed over to the company's in-house architects. They gave the towers' sides a slight taper and sculpted the exposed faces at the tower tops to give them greater definition. That, it seems, was it. Pure cosmetics, lightly applied.

Architects' design input varies from zero to 100 per cent at the concept stage, with zero being much more common. Later on in the process, the architectural design contribution can, however, still be crucial. When French architect Alan Spielmann was commissioned for a new viaduct near Clermont Ferrand – within sight of Eiffel's iconic Garabit rail bridge of a century before – the engineering solution was, once again, an a priori decision. But it took Spielmann a further year to refine the solution – in particular, reducing its visual squatness. Other bridge architects, such as Ronald Yee, who works predominantly in the Far East, are involved sooner. Yee is a staunch proponent of the classical 'golden section' as a proportioning system for bridges, and uses it both vertically and horizontally on his Ting Kau cable-stay bridge and Tsing Ma suspension bridge (fig. 6), both in Hong Kong. At Tsing Ma, even the aerofoil deck sections were golden-proportioned.

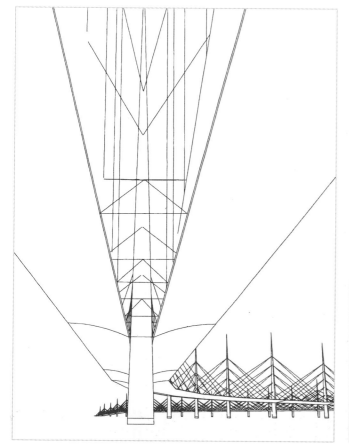

1 2 | 4 5
3 |

figure 1:
The Pont Valentre over the River Lot, Cahors, early fourteenth century.

figure 2:
Tower Bridge, London, 1890, by the architect Sir Horace Jones and the engineer John Wolfe-Barry.

figure 3:
The Wearmouth Bridge, Sunderland, 1796, by the architect John Soane and the engineer Thomas Wilson.

figure 4:
Preliminary scheme for the Grand Viaduc du Millau, Cahors, by Foster and Partners.

figure 5:
William Natcher Bridge, 2002, Owensboro, Kentucky, by Vijay Chandra.

Yee also uses a version of another classical device, entasis, shaping his cable-stay pylons to counteract the optical illusion of bulging caused by the splay of cables.

The current fashion for splitting the decks of wide bridges mirrors the architectural preoccupation with breaking down the bulk of large building complexes. What might be one large building can become a township of smaller structures, bringing the landscape into the design. With bridges, the need for balance tends to provide less scope for complexity, though for example Zaha Hadid's (fig. 7) leaping architectural forms lend themselves particularly well to bridge projects, and she is responsible for several. Perhaps more typical might be Nicholas Grimshaw's bridge at Ijburg (fig. 8) outside Amsterdam, which must carry road, rail and pedestrian traffic side by side to a new island settlement. The architects saw the double enclosing arches of their bridge as rooms, designed the faceted joints of the arches to resemble those of marine crustacea, and split the deck into several thinner strips so as to provide a visual link with the water beneath. They also decided to link the two arches with a bowsprung under-arch at the central linkage point, so creating the effect of a sinusoidal wave. These were very architectural decisions, taken early in the competition process. The competition itself was unusual, because previously The Netherlands had pursued a policy of reclaiming land to form polders contiguous with the mainland: consequently bridges were of little import. At Ijburg, a different policy was adopted: the new settlements – effectively comprising a satellite town for Amsterdam – would be created on a number of linked islands. Water automatically creates the physical boundary, and the bridges provide the necessary markers at the points of entry.

But what form should such markers adopt? By chance I was present when two English architects – both of whom had won competitions to design pedestrian bridges – met each other at a party. Architect A congratulated the younger architect B on a recent victory. 'Very interesting,' he said, 'there are only a certain number of ways to design a bridge. And yours was one of them.'

Quite a put-down. Yet it reflects the undoubted fact that – despite extraordinary technological advances in recent years – engineers and architects perhaps have less freedom in bridge design than might at first appear. You cannot defy the laws of physics, though you can always – as Santiago Calatrava is adept at demonstrating – delay the point at which the load finally meets the ground by transferring it elegantly through sub-structures. Few can do this with the elan of Calatrava. More usually, bridges treated in this way tend to end up overly complex. If there is a rule, it is to pare down the design – which is why so many bridges look like other bridges. It is rare to find a genuinely original solution, but it happens, and a rich crop has emerged in Britain in recent years. For instance, WilkinsonEyre's S-curved (fig. 9), tilting-masted cable-stay pedestrian footbridge in London's docklands – one half of which, just to make the whole thing even more ingenious, rotated open for shipping. Or Hodder Associates' hyperbolic paraboloid pedestrian bridge in Manchester (fig. 10), like a power station's cooling tower placed horizontally, the shape of which allows the footway inside the glazed tube to slope between the different floor levels in the buildings on either side. Or indeed the Royal Victoria Dock footbridge, out east in London's docklands, by architects Lifschutz Davidson and engineers Techniker (led by the author, Matthew Wells). There the concept of a transporter bridge is revived and combined with a dash of structural audacity: a 'Fink Truss' system is inverted, and in the process subverted, to provide a highly articulated profile very necessary in the flatlands of East London.

The question of whether a supposedly pure engineering structure is better or worse than one in which an architect has taken a major role will never be resolved. Particularly in the case of those who, like Santiago Calatrava, are themselves both architect and engineer and

so afford the spectacle of two halves of the same brain in dialogue. I was doubly lucky in being made aware, quite young, of two of the finest 'engineer's bridges' in the United Kingdom. At university in Durham, most days I had to cross the Kingsgate footbridge (fig. 11) over the gorge of the River Wear. Though I did not know it at the time, it was the last project to be personally designed and supervised by the great engineer Ove Arup, from 1962 to 1963. The trough of the concrete footway felt as if it should have been an aqueduct, in the tradition of the Roman Pont du Gard in France, or Thomas Telford's great Georgian canal aqueduct at Pontcysyllte in North Wales, its ironwork superstructure perched lightly high above the valley of the River Dee. Arup's bridge shows you how it is made: two precisely balanced sections built parallel to the river banks that were then rotated into position and linked with a pair of simple, fully exposed, bronze roller expansion joints. It is clearly alive, this bridge, quivering slightly as the gaggles of students career across. But just as interesting is its setting. It joins ancient Durham, with its cathedral and castle on a wooded peninsula, with the modern Durham of new university buildings. It has to leap not just a river, but a thousand years. It achieves this task effortlessly.

Shortly after this, I saw the Humber Bridge (fig. 12) being built in north-east England. By William Brown and others of Freeman Fox and Partners, at the time it was the longest single-span suspension bridge in the world, comfortably outstripping the 1964 Verrazano Narrows bridge across New York harbour that had previously held the title. It is so long, indeed, that its towers had to be angled slightly to allow for the curvature of the earth. Visiting the East Riding of Yorkshire at intervals, I received a stop-motion picture of its progress. First the towers, then the fragile pilot threads, then the cables spun across, and then, what seemed to be the truly miraculous part: sections of deck were floated out into the tideway and hoisted into thin air. The narrow band of the deck advanced from either shore – ribbons in the wind, given the extraordinary height and length of the bridge. The distances were such that the eye was deceived. Cables visually dissolved in the mist of the estuary. The impression was of impossibly slender cantilevers poised high above the waves, as if in a weightless environment. When they finally joined, and the bridge became a bridge, it retained that delicate, evanescent quality. It is an extraordinarily beautiful response to the land, water and sky: very different to the urban setting of the Verrazano Narrows. In more recent times perhaps only the marvellous 1994 cable-stay bridge of the Pont de Normandie in northern France (another great estuary crossing) approaches the atmosphere of the Humber Bridge.

But the Humber crossing came about as a purely political gesture: a government wanted to make employment and gain votes by uniting the industries and road networks of Lincolnshire and Yorkshire, either side of the estuary. The new county of Humberside was created around the bridge. This was an artificial construct that did not last: politically, Humberside has now been abolished. The bridge, opened in 1981, is quite possibly unnecessary anyway, since most traffic in that part of the country continues to run east-west rather than north-south across the Humber. The bridge remains saddled with a huge debt. But it was, and remains, a most potent symbol as well as a technological marvel and a thing of beauty. It is no coincidence that it was conceived by the political generation that was simultaneously responsible for the supersonic jet, Concorde.

No coincidence, either, that its record-breaking characteristics should have in turn been eclipsed by other politically-generated bridges, most notably the series of connections between Denmark and Sweden. Dissing and Weitling's suspension bridge of the Great Belt, opened in 1998, is even longer than the Humber. Attention soon passed from that, however, to the hybrid connection of the Oresund Link, a ten-mile road and rail connection with a combination of bridges and tunnels. The main bridge section (designed by a consortium

including Ove Arup) is at 490 metres (1,608 feet) the longest cable-stay span in the world. It is also the strongest, since it carries both a motorway and, beneath it, a dual-track high-speed railway. This is scarcely a new concept – in 1855 the Prussian-American engineer John Roebling made a double-deck road and railtrack suspension bridge across the gorge of the Niagara Falls, and there have been plenty of other dual-mode bridges since – but the scale and ambition of the Oresund Link places it in a different league. To open it, in July 2000, the Queen of Denmark set off in one train from Copenhagen Central while the King of Sweden departed simultaneously from Malmö Central: the two monarchs, with their prime ministers in tow, met on the artificial island of Pebeholm to mark the ceremonial opening. As with the Humber Bridge – but with greater justification – there is talk of the link creating its own economic zone, neither Sweden nor Denmark, just 'Oresund'. The effect of building it has been compared to the demolition of the Berlin Wall.

There will always be such dreams. There are inevitably plans now to link Denmark more directly with Germany. That fades into insignificance compared to the project to build a bridge across the Straits of Gibraltar, so linking the key nations of the Magreb. It is unarguable that Andalusia has always had more in common with North Africa than with Northern Europe. Looked at coldly, the distance to be spanned is not insuperably great, and the technical problems are relatively straightforward. But politics are never straightforward: there, the desire for economic growth is balanced by the perceived problem of economic migration northwards from poor Africa to affluent Europe. When sea is to be crossed, it can to some extent be controlled. Politicians do not necessarily want all the aspects of free movement that a bridge brings. But setting aside all that, there is possibly only one more symbolically important crossing to be made in the entire world than the linkage of Europe and Africa: the linkage of America and Russia, across the Bering Strait. The economic case in those remote regions is perhaps difficult to make, but plans are progressing anyway. How, when bridges are finally designed in these places, do their designers begin to acknowledge the overwhelming importance of such connections?

Quality of design has normally been high on the agenda of such great national and international projects. The aesthetic falling-off of the post-war period took place elsewhere. It can perhaps be argued that the art of bridge design merely became overlaid and obscured by the proliferation of workaday structures thrown up by a huge expansion of road networks across the world. In road-building, as in the preceding railway and canal ages, the most original or, at least, most aesthetically convincing designs were usually generated by the most dramatic topography rather than by any overt desire for symbolism – the mountain viaducts of central and southern Europe, for instance. New fast motor roads affected bridge design in other ways. Firstly, it became commonplace for a single style – or, at best, a very limited range – of bridge designs to be employed by each highway authority. This created a degree of homogeneity, but also awkward situations where standard designs were distorted to fit all situations. Secondly, the designs had to deliver their aesthetic message, such as it was, at speed. The faster the car is travelling, the narrower the motorist's angle of vision tends to be – as well as the shorter the overall viewing time.

Some few brave bridge designers fought against this. Switzerland's Jörg Schlaich, for instance – famous for highly sculptural road bridges – found that, when it came to designing a humble motorway overbridge at Kirchheim (fig. 13), the expressive structure he suggested, made of tensioned cables and struts, was frowned upon: such designs, it was felt, would invite vandalism and sabotage. So

Schlaich found himself obliged to encase his 1993 structure in concrete, while maintaining the profile he had originally wanted.

Architects do design read-at-speed bridges, but their talent for detailing has so far been more exploited on examples for pedestrians, particularly in urban regeneration projects. There is a colossal difference between the macro-detailing of, say, the Great Belt Bridge in Denmark – which is all about a speed-journey – and the micro-detailing of some of the present crop of pedestrian bridges which are not so much to do with journeying as with promenading. Texture and feel of surfaces, scale of components, sense of movement underfoot, a feeling of shelter and security, even sound and light reflections, become important. A timber deck may feel warmer and more human, but can become slippery when wet: how to keep that human feel when using man-made materials? A consistent curve of the deck, rising to and falling from the point of maximum clearance, looks best – but how to square this with a requirement to provide flat refuge-points for those in wheelchairs? The all-important handrail – could this be made unique or at any rate special, for instance with a coating derived from industrial diamonds as one competition-winning British bridge design proposed?

Such details are very much the preserve of architects rather than highway engineers: human interface is what they grapple with in their work constantly. Thus Norman Foster could have a more significant design input into the Millennium footbridge across the Thames in London from the Tate Modern gallery to St Paul's Steps on the north bank than he could at the Millau viaduct. In London, he required of his engineers, Ove Arup, a structure so slender as to resemble 'a blade of light'. This led to a design that is a highly tensioned suspension structure without towers, the footway sitting directly upon the cables rather than hanging from them. The design is inspired by the spidery rope-bridges of central Asia. Like them, it was inclined at first to swing from side to side, which it did rather more than had been anticipated. Remedial damping work was needed, but the all-important image was carefully maintained – the bridge had become a popular attraction even when closed. In that case, then, an architect created a seemingly reduced, minimalist structure which might at first glance be mistaken for an act of 'pure' engineering – but this was not the case. Architects and engineers worked together in the service of an aesthetic ideal, and while the aesthetic reference was ancient, the means of achieving it was extremely modern: so modern as to require entirely fresh thinking.

Frequently such historical references crop up in architecture, less often in bridge design, though there are contenders. One such is the 1673 Kintai-Kyo bridge in Iwakuni (fig. 14), Japan, which looks startlingly contemporary, being so reduced to essentials. Timber arches between rubble stone piers form an undulating walkway of stairs and smooth crowns: the arches are the walkway, there being no horizontal deck placed above them. The entire timber structure of this bridge used to be renewed every 25 years, by rebuilding one of the five arches every five years. This stopped when modern timber preservatives meant that the quinquennial renewal process could halt – or at least be considerably slowed – following a complete rebuild in 1953. Hence, to some extent, its modern feel. Crossing this bridge may be hard work, but its ceremonial power is immense. Why, after all, should a bridge not undulate, assuming provision is made for the less able-bodied?

Visual reference to the past is also made with the current generation of floating pontoon bridges – Professor Eiichi Watanabe of Kyoto University has proposed a coastal road bridge consisting of a snaking box-girder deck placed on cylindrical floating concrete pontoons, for instance. The Romans would have appreciated the idea. Similarly with the latest ribbon-bridges. Instead of a footway of lively

12 13 | 14 15

figure 12:
The Humber Bridge, 1981, by Freeman Fox and Partners.

figure 13:
The engineer Jörg Schlaich's overbridge at Kircheim, 1993.

figure 14:
Kintai-Kyo Bridge, Iwakuni, Japan, 1673.

figure 15:
Pulteney Bridge, Bath, 1773, by the Adam brothers.

planks as in the past, you have a single narrow band of thin reinforced concrete held in tension. As with so much to do with bridge design, this type of crossing shares technology with the design of roofs – such as Alvaro Siza's virtuoso catenary-curved thin-shell concrete roof at Expo '98 in Lisbon, Portugal. Structurally, large clear-span roofs and long bridges can be near identical: the difference is that the roofs of buildings seldom have to be designed to bear constantly varying loads of traffic as well as wind. Aesthetically, bridges can learn from the buildings comparison. Too little attention is often paid to the visual effect of the undersides of bridges. In buildings, naturally enough, the roof soffit receives a high degree of design attention.

As Matthew Wells recounts in this book, ways of building bridges were refined immensely during the twentieth century – taking great leaps forward during the war years, when technology transfer from other industries such as shipbuilding was at its most intense. New ways with old materials, the arrival of new materials, the increasing importance of aerodynamics, and above all the rapid rise in computer programmes capable of 'testing' designed but unbuilt structures across a range of conditions – all these things led to a point where, from an engineering point of view, bridges could be made to perform

structural gymnastics. What then began to occur – a process that is accelerating today – was a reassessment by both engineering and architectural professions of what bridges meant aesthetically.

In a sense, what has been taking place is a reassessment of the nature of bridgeness. To come back to the start of this introduction: what does a particular bridge set out to achieve? Some examples deny their bridgeness, such as the Adam brothers' Pulteney Bridge (fig. 15) in the Georgian city of Bath, lined as it is with small shops. Being an extruded street, like the medieval London Bridge before it, it is perfectly possible to cross without being aware that the River Avon runs below it at all. Although you might wonder idly why the imposing, tall buildings on either side have given way to delicate, small-scale buildings – lightweight, for the time – your eye is held at street level by the standard-gauge Georgian shop windows, so the pretence is maintained. Such denial of orientation – a charge you could also level at windowless tubular-structure bridges, effectively tunnels in the air – does not apply to the Rialto in Venice (fig. 16). The Rialto offers you both a giveaway steep arch and open-deck pedestrian bypasses to each side of its always crowded central street. Nor does the denial of view occur in other English Palladian bridges of the eighteenth and

early nineteenth centuries, where, although the bridge is treated as a piece of architecture, it is as a room with a view – and serves, in the Palladian manner, to frame and enhance that view, both from within and from without.

These issues received a fresh airing in the 1980s and 1990s with something of a revival of the notion of the 'inhabited bridge', and all these issues came to the fore. Some ideas were really little more than extrusions of the cityscape on either side – culverting the river and denying its existence. A derivative of this later proved successful in mending the wounds through cities made by traffic arteries: the highway is effectively culverted, and the landscape replaced over the top. Architect Piers Gough's 'Green Bridge' (fig. 18) in East London, for instance, unites a park. Other bridges unite educational campuses bisected by roads. Should the road beneath be acknowledged or denied? Some designs now incorporate pause-points, even belvederes, into otherwise enclosed bridges to provide the necessary view. A direct architectural comparison is with the endless extension of terminals at airports, usually involving the construction of covered bridges. Again, the earlier tendency to keep such links fully enclosed and artificially lit has been replaced by a new aesthetic of

transparency. Those who regard inhabited bridges as rare should look at airports with fresh eyes.

At the start of the twenty-first century, bridges have now become accepted symbols of regeneration and revived civic pride, working on the micro-scale to connect parts of the urban fabric. The contrast is with those bridges that work on the macro-scale to connect nations, even continents. It might seem a paradox that, as some bridges have got steadily bigger, others have simultaneously become a great deal smaller, but both are aspects of the same phenomenon. There is, perhaps, a greater emphasis on urbanism, on the space between buildings and districts. It has also been well noted that a bridge, because it facilitates movement, can help to correct the economic gradient between rich and poor sides of town. When a bridge is flung across the dividing line – be it a waterway, road or railway – the effect is almost hydraulic: the money starts to flow from the rich side to the poor side. This is why, in so many urban regeneration projects around the world, the bridge is an essential part of the process.

Again, this works on all scales. At one extreme, you have a modest new bridge to revitalize the part of town – any town – on the wrong side of the tracks. It might be regarded as post-industrial Band-Aid,

since its task is usually to help rescue pockets of economic deprivation left high and dry by the receding tide of blue-collar jobs – but if so, then nobody can deny that the bridge, as an urban form, is peculiarly suited to this task both physically and emotionally. At the other extreme, you have something like the 1998 Jamuna Bridge in Bangladesh: five kilometres (3 miles) long, with another 30 kilometres (18$\frac{1}{2}$ miles) of approach roads, crossing the braided, constantly shifting, earthquake and flood-prone course of the River Jamuna or Brahmaputra. That low, multiple-span bridge is scarcely spectacular apart from its length, but it unites the agricultural west of the country with the industrial east. Two nations become one.

The picture is of course not all rosy. Yes, a bridge can heal divided communities, but then again it can be a focus of sectarian hatred. Bridges are targets for bombers. Bridges are sought out by refugees. A bridge can offer more than it can deliver: a bridge built from nowhere to nowhere – as sometimes happens in anticipation of development – might create a sense of place, or it might remain marooned and ridiculous. A road bridge can also be too successful in sucking in polluting traffic and sprawls of buildings: like all roads, they can generate their own traffic just by virtue of existing. And a bridge,

like any building, can outlive its usefulness and become an artifact of industrial archaeology – a more potent and iconic artifact, however, than many a conventional building, as Thomas Pritchard's original eighteenth-century Iron Bridge (fig. 17) at Coalbrookdale in the long de-industrialized English county of Shropshire bears witness, as do a thousand abandoned railway viaducts all over the world.

Bridges provide a way to understand and appreciate the life of a town, city, country or economic region. Even abandoned and derelict, they imply movement, action, prosperity, power. Bridges make, perhaps, the most evocative of all ruins. The broken twelfth-century bridge of St. Benezet in Avignon – abandoned as a crossing in 1680, so becoming a tourist magnet ever since – gives us the clue. The single most fascinating aspect of today's intense design focus on this area is therefore the beguiling notion of the bridge's function not as journey but as destination, an end in itself.

16 17 | 18

figure 16:
Vincenzo Scamozzi's design for the Ponte di Rialto, Venice, 1588.

figure 17:
The Iron Bridge, Coalbrookdale, 1779, by Thomas Pritchard.

figure 18:
The 'Green Bridge', Mile End, London, 1998, by architect Piers Gough.

Bridge Design: A Brief History Matthew Wells

A critical framework for bridge design

The way today's bridge designers approach their task, and what we see in their work, are essentially predetermined. Our responses are guided, perhaps subconsciously, by an underpinning of received knowledge. We know how bridges should look and we can sense the forces they contain. When we look at any single structure, we bring to it the range of other bridges we have seen. Similarly, the designer's sketch holds within it all the other forms he might have made.

The criteria we apply in order to appraise and understand bridges are seldom consciously chosen. They are not systematic, or even necessarily commensurable, as is evident in the contradictory responses evoked by the work of Spanish bridge designer Santiago Calatrava (1951–), which can inspire passion among architects but give rise to reservations among engineers. The accepted way of looking at bridges at any given time merely represents the most dominant viewpoint, which has overruled other interpretations, and the rules we apply when assessing structures are adjusted to the continually shifting themes of design.

It is too simplistic to characterize the changing face of bridge design as continuous technical advance or as a dawning of aesthetic awareness. Developments in bridge design proceed as a sequence of unique events through the realization of individual projects. Change has occurred across a broad front, often by accident or incidentally, and many viable lines of advance have never been taken up. To grasp our present position, it is helpful to look back through the layers and accretions of knowledge to find the points where current ideas first surfaced and were adopted, the threshold developments and the paradigm shifts when successive understandings of bridge design were replaced or radically recentered.

Ancient bridge-building: geometry and power structures

Several types of bridge structure have come down to us from ancient civilizations, including timber-corbelled bridges formed from inverted stacks of logs, which still exist in China and eastern Asia and have no modern counterpart. The 'Rainbow bridges' of the Chinese Han empire (206BC–AD220) – arches made up of interlocking log segments for short crossings over canals – have disappeared except in illustrated form. These types are unequivocally similar to Roman examples, providing evidence of communication and trade, or perhaps correspondences, between opposite ends of the ancient world.

It was Han engineers who advanced the suspension bridge into the form we recognize today. In the first century AD the dynasty was securing its western border through a policy of appeasement in which neighbouring hordes were kept at bay by a process of bribery and assimilation. More than a third of the empire's economic output was made up of luxury goods to be transported west. As a result canals were developed and fibre tow ropes refined to consistent qualities and strengths. These cables could, in turn, be used in suspension bridges to extend the road network in areas difficult to cross. Multiple-span suspension bridges, lightweight, economic and without the large foundation forces associated with the arched alternatives, suited the wide rivers' flood plains with their weak alluvial soils.

As diplomatic relations gave way to war the Chinese reaffirmed their technological superiority in more traditional areas. Advances in metallurgy, made to improve swords and cannon, also gave them the high-strength forged-metal chains that directly replaced parts in some of the empire's strategic organic suspension bridges.

Ancient examples of masonry arch bridges exist in both the East and the West. The first examples to be manipulated in a recognizable form date from as early as 1000BC and were first built in Persia, where

mud-plastered bent reeds produced small vaulted huts. The free profile of bent stems tied across the top approximates to the ideal parabolic profile for a free-standing arch, making it a classic example of pre-adaptation, a structural advance developed out of a form generated by other influences. This arch form spread through the Near East, and can be seen in the Turkish bridge illustrated here (fig. 1).

Approximately 500 years later the Romans were using semicircular arches, which are easy to adapt to different requirements and amenable to prefabrication. Although less efficient than the segmental arches (smaller fragments of circles) of ancient China, these semicircular arches were built to a greater size using the Roman invention of concrete as a core behind close-jointed stone arch rings. Across the deserts to the east of the Roman Empire the Sassanids of Persia employed captured Roman engineers to build burnt-brick semicircular arches, dispensing with their local forms of parabolic arches despite the fact that they were intrinsically better. A power structure based on fear of the enemy was therefore able to overturn a superior technical logic.

The huge colonnade aqueducts of the Augustan age (27BC–AD14) propagated the permanence and power of the *Pax Romanum*, sending out a message that the imperial world would last forever.

Transport and other infrastructure projects became symbols of the shift to a new order based on trade, but Roman engineering continued to embody a tension between the Empire's desire for expansion and its need for continual consolidation.

Strategies involving fortifications and siege works were central to the Roman army's success. Demountable timber truss bridges (fig. 2) are described by the Roman architect and theorist Vitruvius in his treatise on architecture dedicated to Augustus, the only such work to survive from antiquity. These temporary structures kept the army moving effectively and, in the early days of the Republic, in the time of Horatius, the removal of the Tiber bridges was an important component of the city of Rome's defence strategy. (The constituent timber was too valuable to burn.) Contemporary religion reflected deep-seated popular concerns and forbade the fixing together of timber bridges with metal cramps of any kind. This tension between the reassurance of repose and permanence on the one hand, and faith in the ephemeral and temporal nature of a lightweight bridge on the other, continues to colour our responses today.

Until the work of one individual, the Italian astronomer and natural philosopher Galileo Galilei (1564–1642), established an algebraic basis for engineering design, ancient and medieval bridge-building in Europe was firmly grounded in the geometry employed by masons and carpenters.

A strong predisposition towards geometric tools for the description and sizing of structural elements was grafted onto a deep-rooted craft tradition, which prescribed how materials should be handled. Setting out systems could be learned (and therefore protected), while proportions could be safely extrapolated through sequences of similar structures. Methods of measurement and construction using rules devised by the Greek mathematician Euclid were, however, mingled with esoteric and arcane ideas. The extent to which rules of thumb were applied to bridge structures is difficult to determine in the absence of written documentation, and the evidence in the stones themselves waits on modern archaeologists making representative samples of earlier engineering and statistical analyses.

The evolution of forms by selection is a verifiable mechanism of ancient times. Medieval bridges often show accretion and documents of collapses accurately describe failure modes and rational counter-measures in subsequent rebuilds. Many very early structures achieve levels of efficiency that seem extraordinary to modern practice. But

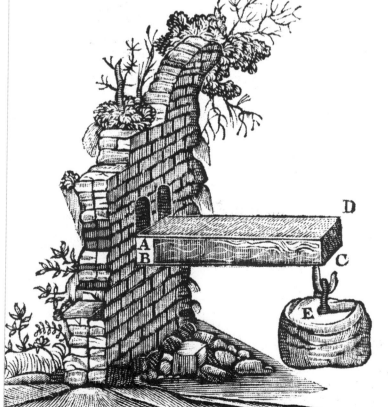

figure 1:
The Seljuk-era bridge over the Eurymedon River, Turkey, eleventh century. The parabolic arches were an ancient form developed from the bent reed buildings of the region's marshes.

figure 2:
Roman timber truss bridge across the Danube, second century AD.

figure 3:
The idealization of a cantilever as studied by Galileo Galilei. Having abstracted a general condition from a specific case, he applied Archimedes' principle of levers to work out how strong to make the structural member for a given load.

the mental devices used to size vaults, roof trusses and other masonry structures may have had only limited application to bridge-building, a job which does not appear to have had celebrated masters in the Middle Ages. An echo of these geometric systems reverberates in the funicular lines adopted by Victorian arch builders and in the simple rules of thumb and proportioning systems that all modern engineers rely on to develop preliminary schemes and to check finished projects by comparison.

1620–1750 The rationalists: conceptual bases are constructed
The idea that all structural engineering, and therefore bridge-building, can be treated as a deterministic science can be traced to one man. We assign to Galileo the decisive step towards our modern understanding, although his insights were, understandably, influenced by precursors. Before a single body of engineering knowledge could be condensed there were many engineers developing just enough theory to be of immediate use in their work.

One such individual was the Dutchman Simon Stevin (1548–1620). Intensely pragmatic and highly numerate he is famous for having flooded the polders during the Dutch Revolt (1565–81) to confound the Spanish. His engineering consciousness was on the cusp between medieval and modern. Schooled in the writings of the Greek mathematician Archimedes he was able to see beyond the principle of the lever and devised the parallelogram of forces: the idea that one force can be augmented and redirected by another. This apparently obvious notion – half geometry, half vector algebra – forms the basis of our intuitive grasp of forces acting within form. He invented a proto-calculus derived from the problems he had met, which he then reapplied to other cases. The influence of his approach on the later science-based formulations of calculus has ensured that method's seemingly remarkable applicability to engineering problems.

Unlike a modern researcher, Stevin did not explore or extrapolate from the internal logic of the mathematics that he conjured up. He described situations by setting up mathematical models, gave these models mechanisms that 'saved the appearances' (in other words, explained and predicted phenomena observable in nature) and then, still with some degree of risk, predicted future outcomes. This method of using mathematical models, reduced to perceived essentials and checked by adequacy of performance, is universal in modern engineering.

In 1636 Galileo was under house arrest, having defied the Catholic Church by staying true to an innate rationalism and refusing to pass over evidence that the solar system is heliocentric rather than geocentric. In the second of the three books that he wrote during his three-year confinement he sought to lay down a new and general science of structures, an exact and reasoned way of determining the size of structural elements for a given set of conditions. Many basic premises were already in place and enabled him to do this, including the Renaissance belief in complete and encompassing systems, the notion of universality – that whatever works in one place will work throughout the earth and heavens – and the recognition that a simplifying assumption does not necessarily invalidate the resulting solution.

One specific mechanism that Galileo examined was the cantilever, which had presented him with a practical problem during his time at the Venice arsenal. The size of ships' ribs at the arsenal was limited by their temporary, partially supported condition in the shipyard, where they were fixed to the keel but not yet planked in. Galileo abstracted this general problem to a standard case, and his famous diagram (fig. 3) which sets it out is a telling hybrid, illustrating the premise but reducing the problem to a small number of measurable and controllable parameters. Following his procedure of simplification he tried to apply geometry to solve the problem, almost a form of the graphical analysis of forces that masons used in the preceding century.

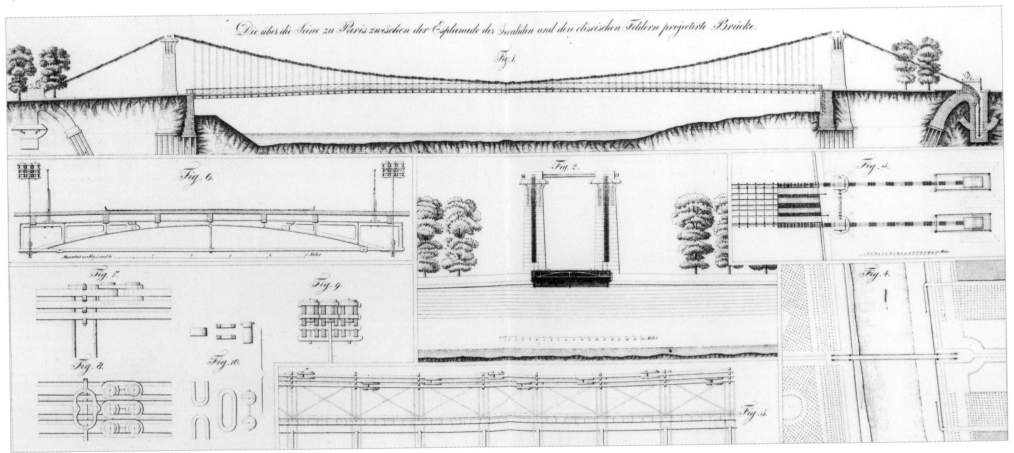

Unfortunately he knew too little about material behaviour, thinking that substances would tear before being crushed, and so he placed the fulcrum on which his cantilever turned at the base of the section and not in its correct location at the centre. It was only in 1713 that a correct solution was found, by the French mathematician Antoine Parent, who was guided by his use of calculus. Despite the imperfection of Galileo's results, however, his conceptual approach, shaped by his geometric understanding, became universal practice.

The stage was set for the development of a particular relationship between theory and practice that is still maintained today, in which neither leads but each influences the other. Throughout most of the late seventeenth and the eighteenth century, construction remained centered in the craft tradition, while more and more conceptual tools were accumulated. The English mathematician Robert Hooke (1635–1703) studied the behaviour of real materials and mathematically idealized them, making elasticity the preferred and readily tractable description of how structures deform. The English mathematician Isaac Newton (1642–1727) and the Dutch philosopher Gottfried Leibniz (1646–1716) completed the discovery, or perhaps construction, of calculus as the mathematical technique most capable of describing the continuous changes of nature.

1750–1850 Early modern engineering: an establishment is formed
The acceptance of mathematics as an essential tool of technical design and development is a European sensibility given final form by the French in the years of Napoleon Bonaparte and the revolutions. The contributions of Claude-Louis-Marie Navier (1785–1836) epitomize this system, whose development was based on merit while retaining a professional protectionism in engineering. Born in Dijon, the son of a well-to-do lawyer, he was precocious at mathematics. Following the death of his father, when he was 14, he went to live with his uncle, Emiland Gauthey (1732–1807), a well-known engineer of the old school. The old man did much to temper his nephew's theoretical bent with practical experience. Navier edited and republished Gauthey's textbook on bridges and channels after his death, supplementing it with notes on the theory of elastic bodies and the bending of prismatic bars. However, his mathematical excursions proved of little immediate practical use and some of the solutions were in fact based on wrong assumptions and therefore incorrect.

He appears to have been unaware of the works of Charles Coulomb (1736–1806), a Frenchman who had published correct conclusions to similar problems. Coulomb had spent long years as an engineer on the colonial island of Martinique, virtually as an exile,

where he had plenty of time between practical installations to develop appropriate and pragmatic theory. His memoir of 1773 took over 40 years to be fully recognized but has since become the accepted model for the modern engineer's methodology. Traits of very much older geometric approaches are melded seamlessly in this work with the French mathematician's favourite tool, calculus.

Navier's own book begins with a history of bridges set out as a process of rational development. On the basis of this text he was sent by the French government to report on English practice. During two visits, in 1821 and 1823, he encountered the antithesis of his own theoretical approach to bridge-building, the pragmatic Thomas Telford (1757–1834), a self-taught Scottish engineer, who was then on site with his ground-breaking suspension bridge across the Menai Straits, in Wales, which was constructed between 1820 and 1826.

Appointed Professor of Calculus at the Ecole Polytechnique in Paris on his return, Navier published another textbook on the strength of materials. It contained the first true evolution of a general method of analysing indeterminate structures. Through addressing this issue the nature of large bridge design changed fundamentally.

Indeterminacy is the property of a structure that has more than one load path. Forces run down through structures away from the

figure 4:
The Pont des Invalides Paris, 1824–6, by Claude Navier: an early suspension bridge design of great sophistication. Construction was abandoned following foundation problems and a focused political campaign.

figure 5:
Thomas Telford's suspension bridge across the Menai Straits, 1826. The high air draft allowed the passage of the Admiralty's sailing ships. Despite the heavy catenaries the slender timber deck experienced aerodynamic problems.

load and out to the supports. If the overall load is shared by two or more routes then a distribution must be worked out, which can only be done by assessing the stiffness of the various parts. More rigid elements resist greater forces with less deflection and therefore attract load towards themselves. Until proper methods had been developed to deal with this complexity it was much safer for engineers to make structures determinate: simple enough or with strategically placed details so that only one set of load paths could exist and the forces were then immediately known. These expediencies obviously limited the possibilities of form.

At the same time as making these theoretical studies Navier embarked on the construction of a bridge, the Pont des Invalides in Paris (1824–6; fig. 4). The structure was always intended to be an exemplar of suspension-bridge design and all the elements were sized from theoretical principles. Construction began but problems soon arose because the cable anchorages were initially undersized and slipped in the alluvium bed of the River Seine. Furthermore, the linked suspension bars were grouped and, by shadowing one another, expanded unevenly in the sunlight. Construction of the partially completed bridge was cancelled and the structure was dismantled in 1827.

Meanwhile, Thomas Telford completed the Menai Straits Bridge (fig. 5), a slightly longer span with a considerable air draft to allow the British Admiralty's sailing ships to pass underneath. His design was predicated on a much wider general experience, which subsequently earned him the accolade 'the father of civil engineering', and reflected the influence of an earlier work in North America by the Scottish émigré James Finley (1770–1828). The bridge helped to solve a specific problem that had been pressing for 50 years, the opening of a route to Ireland.

Despite the failure of the Pont des Invalides, Navier's work was not in vain. When a party of Hungarians came to England in 1832 seeking expertise for the first permanent Danube crossing between Buda and Pest, Telford recommended his protege William Thierney Clark (1783–1852). He produced a chain bridge of unsurpassed elegance with a catenary curve reduced to the proportions that Navier had made accessible. In response to problems of wind-induced oscillations an early type of deck-stiffening system was added to this structure, which Telford and another English engineer, Captain Samuel Brown (1776–1852), had formulated in their projects.

Nineteenth-century French practice developed most of the theoretical tools then adopted by the world's engineers, but the prevalent modes of education and visualization among bridge engineers today are based on the work of just one man, Gaspard Monge (1746–1818). A good mathematician, he trained as a draughtsman and found an outlet in this for his military aspirations, since his humble birth prevented him from attending the officer training school at Mezières.

Monge's work involved setting out fortifications, at a time when advances in artillery were generating vast and outstanding plans in which the visibility of one point from another was of paramount importance. He progressed rapidly and, in the rationalist milieu immediately preceding the Revolution of 1789, developed the modern form of descriptive geometry through which any point in space can be described by mapping into three perpendicular planes. His summary, the vastly influential textbook *Descriptive Geometry*, was published in 1795.

The system was well suited to industrialized production because, if the instructions encoded in the picture planes are meticulously followed, the result is inevitably an exact copy of the designer's intention. Some skill is needed to visualize complex forms straight from the drawings and, more importantly, the form predisposes designers to work across the face of the drawing rather than into its field. It appears that the system's initial widespread acceptance

throughout mainland Europe, occurring before the advantages mentioned above became apparent, had much to do with France's political power and military success.

The development of orthogonal projection in the geometry of warfare made a radical departure from the alternatives then available. The Reverend William Farish (1759–1837), a Professor of Experimental Philosophy at Cambridge University, extrapolated from artists' theories of pictorial perspective, themselves rationalist attempts to define the real world systematically, to produce and promote isometric perspective. This is not in fact a true perspective projection but rather a good compromise of pictorial representation with measurable dimensions, by which the general disposition of a complex assembly can be readily assimilated. Three-dimensional representations are only now returning to importance in engineering design. Other forms such as proportional systems, encoding devices that could incorporate design information, became obsolete following Monge's innovations because of their opacity at a time when the earlier preference for concealment was being overturned by a new spirit of openness and a desire for transparency of working.

Monge, and his contempories with similar skills and backgrounds, were, however, most influential in another way. The Revolution of 1789

brought them to the fore and they seized the day. The old French schools were discontinued while, at the same time, France was at war with a European coalition and numerous engineers were needed to build fortifications, roads and bridges. Monge proposed a universal school to the revolutionary government, with admission based on merit rather than class. The idea was approved and instruction began in 1794. The name 'Ecole Polytechnique' appeared a year later.

The school was a success for three specific reasons. Admission was by competitive examination, not only making it a meritocracy but also ensuring a common level of entrance. In addition, the conditions in France meant that many outstanding scientists, mathematicians and professors were idle in Paris and available for teaching, and the original staff was therefore immediately of the highest quality.

Thirdly, the nature of the teaching was revolutionary, as can be seen through a comparison with what went before. At the earlier Ecoles students had been taught under an atelier system, in which small groups or individuals were instructed by practical engineers in how to design and construct a particular structure. More fundamental mathematical knowledge was added to this core in individual lectures by professors or older students. This faux apprenticeship system was overrun by the new ideas. The many branches of engineering were

taken to have a common theoretical base in mathematics, physics and mechanics. Students followed complete programmes of lectures in these fundamentals for two years before progressing onto condensed engineering courses in a third year. Practical engineering was then learned by moving to a higher institution, the Ecoles des Ponts et Chaussées or Ecoles des Mines, for instance. The morale of the school was boosted by Monge's idea of arranging additional lectures from leading practitioners, whose enthusiasm proved infectious.

The school turned out good if rather academic engineers, who were able to approach new problems with a range of conceptual tools. Ideas could be taken from one area of knowledge and applied elsewhere. But, in addition, it was always possible to stay in the theoretical and academic research trajectory within the school. As well as engineers the school began to produce a large number of academic reseachers and professors, propagating a precise set of attitudes more and more widely.

The success of the school led to similar programmes being taken up elsewhere, such as at the Polytechnical Institute in Vienna and the Polytechnical Institute in Zurich. North American enthusiasm led to the Polytechnic's ideas appearing in the foundation of West Point, the military academy in New York state.

figure 6:
Roebling's Niagara Suspension bridge, 1850, illustrates the key innovations he made in suspension bridge design: wire rope cables and a deep stiffening girder. When the tie-downs shown in the photograph were removed in 1864 to allow an ice floe to pass underneath, the structure collapsed in a gale.

figure 7:
The Menai Britannia Tubular bridge, 1850, by Robert Stephenson displays an immensely strong and graceful box girder. Problems of local buckling and fabrication were rigorously identified and resolved as the design proceeded.

The body of knowledge and the method of teaching therefore became standardized. To complete the process, the codification and communication of knowledge and its advances took set forms. The textbook was introduced as a standard format for exposition, initially made up of collections of lectures, which became necessary in order to teach very large groups. The course-like arrangement of material became general, with initial introductory chapters setting out fundamentals and the following sections building up into a complete survey. An end-piece directed the reader to further reading and the latest innovations.

Industrial exhibitions also began to appear during this period, the first held in Paris in 1798. Partly demonstrations of nationalism, these expositions juxtaposed wide ranges of artifacts, allowing more people to recognize ideas and transfer them across technologies. The importance of simply seeing what a new invention looked like should not be underestimated.

The establishment of an academic superstructure over the practice of engineering had far-reaching consequences. Advances were presented through papers and journals vetted by peers for authenticity and veracity. The overall quality of output was therefore assured and, at the same time, a body of received knowledge was confirmed and then entrenched. The academic codifications of the Napoleonic French consolidated the notion of generality, that methods should be of universal applicability to be useful and safe. The belief that general principles should inform each particular case has shaped the development of structural form ever since.

Early modern engineering: a heroic age

Construction in the years from the 1850s to the early twentieth century was characterized not only by the widespread introduction of steel, followed by reinforced concrete and the industrialization of building, but also by a universal idealism in technology. We can only have a slight awareness now of the strength of this belief in applied science, but if there is a universal, innate image of the engineer it is of a self-assured Victorian with a satchel of cigars and a stovepipe hat.

This idealism was the obverse of economic forces. Developments not only answered demand but also became altruistic enquiries. What were the limits of the new materials and how could they be used in new ways? The current idea that forming bridges in the new plastics, a cutting-edge technology, should automatically be appropriate is a vestige of that time.

The cultural differences in engineering practice that set North America apart were largely defined in the second half of the nineteenth century and are only now beginning to fade with the communications revolution. The end of the Civil War, in 1865, released a generation of transport and bridge engineers, since more than 60 per cent of the two sides' manpower had been mobilized in the conflict. Coinciding with this flood of practical engineering expertise was the rail and road expansion that resulted from the countries' reunion.

Bridge systems were transformed in the wave of rebuilding and expansion. The country's earliest bridges had been made from stone, or more often timber. These wooden trestle and truss bridges, made of baulks joined into triangulated patterns, were essentially carpenters' extrapolations of roof frames, which had appeared in pattern books since the Roman period. By the 1830s blacksmithed wrought-iron ties were occasionally being added to improve their reliability.

Metal trusses began to appear after the Civil War in a wide range of configurations, spurred on by North American patent law. Initially these were variations on wooden prototypes, but large enough for the next scale of crossings that was being called for. The simple stick assemblies had the advantage that they could be prefabricated, boxed

and sent down the line to an advancing railhead. In such structures forces run in a straightforward way, axially along each bar, and could be represented in simple diagrams, making analysis easy. These geometric representations of forces only closed if correctly executed, which was an important safety feature in their widespread adoption. Less advantageous was the vulnerability of trusses to detailing error: not only could the inadequacy of any single pin lead to catastrophy, but as a box of bits in inexpert hands they could also be assembled incorrectly. A German émigré, Albert Fink, had patented a very simple and foolproof assembly in 1851 that could not be put together wrongly, but the system was not particularly efficient and therefore its use was short-lived.

Material quality controls came under pressure in the boom years in the second half of the nineteenth century, which led to a series of failures culminating in the Ohio Ashtalabula Bridge disaster of December 1876. The accident was widely reported and soon became a shared memory, representing the worst of rail disasters. It happened when, on a snowy night, the bridge broke under a train. The front locomotive put on steam and climbed out but the wooden carriages behind ran one by one into the ravine, where the little coal stoves used for heating the compartments quickly ignited the wreckage. It

transpired that a derailment on the icy track had initiated the disaster but the bridge design was considered to be suspect and, as a consequence, the engineer responsible committed suicide.

Following the discovery that the wrought iron within the structure was seriously sub-standard, there was a complete upgrade and codification of the ludicrously overdue federal controls on design and construction quality. Development proceeded rapidly but on a more rational basis as the tide of economic development continued towards the south and west. The southern states, with their wide meandering rivers, became a focus for the development of steel truss bridges. These inexpensive structures, strong enough for railways and long lasting in the benign climate, have since become evocative of their part of North America.

The most famous bridge engineer to survive the American Civil War was John Roebling (1806–69) who, working with his son Washington, defined the modern suspension bridge. An émigré from Thuringia in Germany, his innovations seem to have come from maintaining the closest possible contact with the problems of building. He combined a theoretical grounding from the Berlin Bauakademie with knowledge gained using wire rope in the mines of Freiberg, Saxony. He also had experience of the practical, 'theoryless', approach

to suspension-bridge construction of James Finley, gained in the Union Army.

Roebling's first bridge project was the wooden suspension aqueduct (1845) across the Allegheny River at Pittsburgh. Rather than using the familiar chains, he employed brass suspension cables, an early and expensive acknowledgement of the corrosion problems that still plague suspension bridges today. But his breathtaking insight was to recognize the central importance of controlling the relative stiffnesses of the supporting cable system and the supported deck. In an aqueduct the water channel has an infinite intrinsic stiffness because there is no moving, live load to distribute. In contrast, on a railway bridge, the train acts as a huge point load pulling the draped cable out of shape. On Roebling's next project, for a railway bridge above Niagara Falls (fig. 6), he introduced a deep stiffening girder, which became the standard solution for decades to come. The main cables were also stiffened by secondary fan cables to make a hybrid suspension system. Photographs show additional ties that run back down into the river gorge to reduce movement.

Other innovations followed: on the Cincinnati Bridge (1866) the main span was balanced by suspended side spans for the first time. As the cables became bigger on each project the Roeblings turned to

8 9 | 10

figure 8:
Brunel's brickwork arches at Goring, 1846. The arch crowns were made extremely shallow and yet carry the heaviest of modern locomotives.

figure 9:
The series of cheap and temporary timber viaducts for the South Devon and Cornwall railways, 1849–64, was seen by Brunel as a chance to experiment with configurations and details.

figure 10:
The St Louis Bridge by James Eads, 1867, under construction. The triangulated temporary works above the permanent steel arches were forced on the builders by hostile ferrymen who insisted on a full steerage being maintained.

'air spinning', running individual wires backwards and forwards over the crossing to build up a much larger stranded cable in situ.

Their masterpiece, however, is the Brooklyn Bridge in New York. Although John Roebling died in an accident surveying the site, his son continued to pursue the project alone. When halfway through construction Washington found that he had been given sub-standard material by an unscrupulous supplier, he decided not to turn back and undo the completed work but instead wove in more strands than originally intended to complete a structure of adequate strength. The bridge remains a celebrated landmark of New York and the country's first designated national monument.

At the same time, engineers in Britain were creating other discourses. The naval captain Samuel Brown had drawn on his knowledge of ships' rigging to build and patent several very light and elegant eye-bar suspension bridges, in which the main suspension chains are made with long, forged links. Unfortunately, when he came to build a railway bridge he did not foresee the need for deck stiffening and, as a result, trains crossing his Tees Bridge (1830) set up a travelling distortion like a wave in a beaten carpet. The structure was therefore demolished and no railway suspension bridge has since been built in Britain.

The question of which structure can claim to be the first modern bridge is still a matter of debate. Fritz Leonhardt (1909–99), the pioneer bridge designer, believes the accolade should go to the Severn Bridge in England, because of the evident influence of technical advances on its form, but the historian of modern architecture Sigfried Giedion favours the Salginotobel Bridge in Switzerland by Robert Maillart (1872–1940), because of the sensibilities it demonstrates. For me, the Britannia Tubular Bridge across the Menai Straits by Robert Stephenson (1803–59; fig. 7) has a good case to put. The way in which the design evolved and was followed up is contemporary in all its aspects, while the simplicity of the form, directness of the solution and reliance on proportion for appearance are timeless.

Stephenson was a bluff entrepreneur and practical engineer with a background in developing steam locomotion and the railways. He had seen Brown's Tees Bridge and therefore in his design of the London to Holyhead railway he approached the crossings at Conway and the Menai Straits in a new way. Stephenson recognized that although sufficient strength could be readily achieved in a bridge, the control of the overall stiffness and the structure's tendency to deflect was critical to success. He appointed the engineering manufacturer Sir William Fairbairn (1789–1877) as consultant, because of his experience

in building deep-section metal girders, and he in turn sought assistance from the mathematician Eaton Hodgkinson (1789–1861).

The Menai design was constrained by an Admiralty requirement for a large air draft beneath the bridge, and an arch structure was therefore impossible. Brown's debacle disuaded the three engineers from pursuing a suspension system and they began experimenting instead on large-scale model tests of tubular structures. In the process they identified and completed much work on the problem of thin-wall buckling, typically observed in the way a soft drink can folds up when crushed. They made the top and bottom flanges of the tube cellular and investigated various patterns of plates. The tubes were built on the bank and then floated out to be lifted into place with steam-powered hydraulic rams, Stephenson's beloved technology.

The dimensions of the elements were set to make the structure efficient. Once the initial proposition was determined then the consequential problems were resolved systematically, but research was required in virtually every area of the undertaking. The mechanical engineer Richard Roberts (1789–1864) designed a punching machine to prepare the plates for riveting that was controlled by paper cards, imitating the jacquard system invented to run looms automatically. This work has been cited as a precursor of automated manufacture,

years ahead of the electronics needed for its exploitation. It should, in fact, be recognized as a parallel to subsequent development but worked out within the limits of its technological context.

Stephenson's competitor Isambard Kingdom Brunel (1806–59) was another generalist who applied close observation and inventiveness to bridge-building. Born to a French father, the famous engineer Marc Brunel, and an English mother, his work combined pragmatism with flamboyance. He was a tireless self-publicist and his structures exhibited that expression often termed 'daring'. Where the Great Western Railway crosses the Thames at Goring (fig. 8) he built an arch bridge in brick that is very wide and has a beautiful, low, elliptical profile. The crown is so attenuated that contemporary critics refused to accept that it would work, but it remains in use and now carries locomotives weighing five times as much as those for which it was originally designed.

On subsequent railway projects, the South Devon and Cornwall lines, Brunel was faced with the serious undercapitalization of the late railway boom. The deep, wooded valleys and moorlands of the south-west peninsular of England generated 64 viaducts made of timber, using large baulks of Baltic pine assembled in simple patterns and stressed together with tie rods (fig. 9). At the same time

as trying out the widest range of configurations, he simultaneously and progressively rationalized his jointing systems, observing and recording the long-term behaviour of wedged and bolted assemblies under full-size test conditions. The supporting piers were set at close intervals to suit the log sizes and so that masonry arches could be added when money became available. Unfortunately it never did and the surreal ivy-clad columns are still standing to form part of the Cornish landscape.

The precociousness of Brunel's talent shows in the Cornwall railways crossing of the Tamar, started in the spring of 1853. Brunel had been sent a model of a lenticular truss by a Hanoverian engineer, Georg Friedrich Leves (1788–1864), in 1838. This lens-shaped arrangement of arched top boom and catenary bottom tie linked by braces, a Von Pauli truss, approaches a theoretical minimum weight form. Without any mathematical tools of optimization Brunel intuitively refined the design with an oval-section top boom, developing Stephenson's box girder concept, to make a bridge expressing pure structural form.

Brunel's ability to range across engineering problems, transport systems and structures stems from the generality of engineering method European engineers had developed. Designers, who operated

independently but were strongly influenced by contractors, also seemed able to examine solutions with little interference from the surrounding power systems. In North America at this time, the influence of entrepreneurs became dominant. On the one hand unregulated disputes between opposing factions stunted many good initiatives; on the other hand success was the only requirement needed for innovation.

By the mid-nineteenth century, that most remarkable of engineering materials, steel, was becoming abundantly available. Developed for rails, guns and the high-pressure boilers of steamships and locomotives, it was the demands of the Crimean War (1853–6) that spurred on the Englishman Henry Bessemer to patent his cheap steel-making process in 1856. Thereafter, steel became readily available in sufficient quantity and of the right quality and price to be used in large structures. It comprises brittle iron adulterated with carbon to make it strong and resilient, albeit rather heavy, and has a straightforward linear strain behaviour. If overstressed, steel yields but keeps on working, shedding load into other components, a very attractive property of structural forgiveness. (This 'elastic' characteristic is easy to work with mathematically and was therefore adopted rather sweepingly as the behaviour model of all structural materials.)

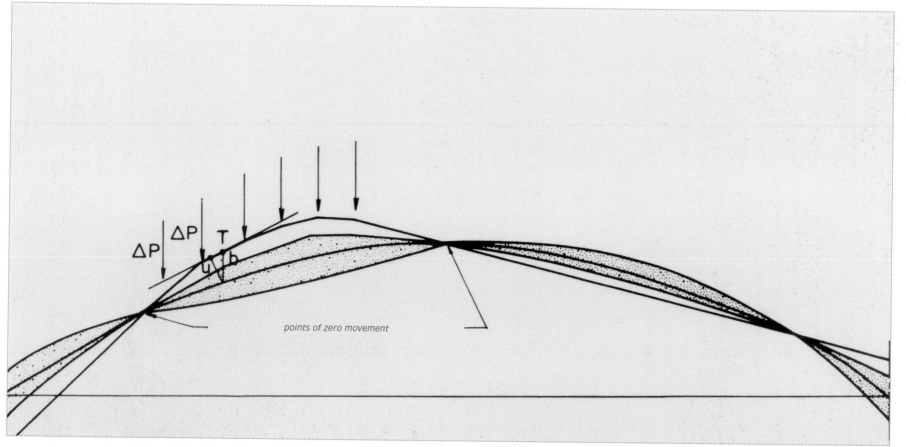

points of zero movement

The Tavanasa Bridge over the Rhine at Grisons, Switzerland, 1905, by Robert Maillart: a small bridge perfectly attuned to its surroundings and to its chosen form, the three-pinned arch.

figure 12:
A diagram from an 1875 textbook by Carl Cullmann, Maillart's teacher. The graphic statics were transcribed by Maillart directly into form.

James Eads (1820–87) was a key figure in the early use of steel in North American bridge-building. He started his career salvaging from the Mississippi River around St Louis, standing in a diving bell up to his waist in the shifting sands of the river bed. Then, during the Civil War, he turned to building the new iron-clad gunboats, learning about the material and how it works.

At the end of the war St Louis was faced with economic stagnation, compounded by its lack of infrastructure: rail cargoes had to be double-handled – transferred from train to boat and then back to train transport – in order to cross the river. In the face of extreme resistance from the boatmen, a bridge project was developed to ease this problem. Eads had never built a bridge before and was therefore obliged to tackle the problem of founding in the treacherous river bottom without any prior experience. Despite this, he proved himself to be a natural expert. His lack of received knowledge seemed to free him to respond in innovative and effective ways to the objections and obstructions he met at every step. The boat-owning lobby called for navigation to be maintained at all times, and temporary formwork beneath the proposed arches was therefore impossible. Instead a superstructure of temporary ties was added and, for the first time, arch ribs were cantilevered out to meet at their centre (fig. 10). Force

patterns in such structures change drastically as the centre connection is made, and Eads was in New York when he received a midnight telegram from the site engineer Theodore Cooper saying that the completed arches appeared to be rupturing. He kept calm, reasoning that the temporary works still partially in place were destabilizing the assembly. It took a cool head to remove some of the structure at that point but his decision proved to be correct.

More prosaically it was Eads's stubborness in demanding unprecedented quality levels in the steel used that was his greatest contribution to the long-term development of North America's bridges. The supplier, Andrew Carnegie, grumbled that it would have been cheaper to make the structure from silver rather than the grade of steel called for by Eads.

The way in which steel was taken up by bridge builders illustrates a common tendency. Other more capitalized industries direct the development of materials science, which is then appropriated by civil engineers. This displacement is significant in the way technologies are applied to bridge-building. The economics of civil engineering structures requires high capital costs for low direct return and so high-strength, long-life, low-maintenance, cheap materials in bulk are the ones that are sought after. By the time a new material becomes viable

and economic enough for use in a bridge its use has already been defined by another branch of engineering.

A confluence of technical expertise and business acumen also appeared in Europe in the last decades of the nineteenth century. The French theoretical background gave engineers such as Gustav Eiffel (1832–1923) and Paul Joseph Bodins (1848–1926) confidence to push arches made from steel to unprecedented sizes. Eiffel's projects were won in competitive tender and provide a distinct contrast to today's practice, where risk appears to reduce many design-and-build projects to abject banality. Eiffel, on the other hand, was able to produce a sequence of very light, elegant gantry bridges set on two pinned sickle arches. These naturally proportioned parabolas are statically indeterminate: internal stresses cannot be found by arithmetic alone. They are efficient in material use but secondary forces, thermal straining and foundation settlements become important considerations. Eiffel built carefully and systematically on his practical experience and so, for instance, the laced struts in his Garabit viaduct in the Massif Central (1884) reappear in the members of the Paris tower.

The reinforcement of concrete (fired, pulverized earth that hardens to artificial rock when hydrated with water) with metal rods was a

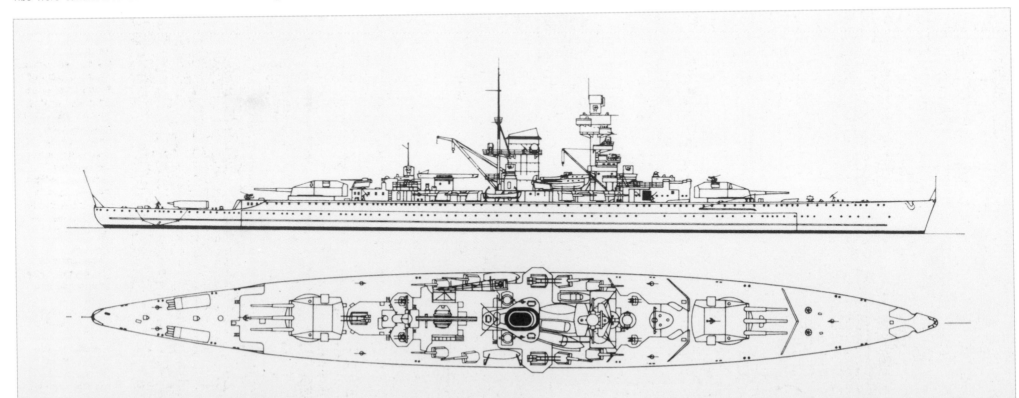

French invention of the 1860s. First used in giant flowerpots shown at the Paris exposition of 1861 its use was rapidly expanded to take in floors, walls, thin shell roofs and building frames.

Technologies develop and mature at different rates and, similarly, economic forces fuel progress to a varied extent. Some systems are more tractable and stable than others. Steel, for instance, was easily improved. Because its quality is readily controllable its behaviour is amenable to simple analysis. Concrete is less stable, yielding up its problems sequentially and demonstrating complex and subtle nuances, all in a material whose perceived cheapness (although it is not ubiquitous) leaves little surplus for research investment.

Major experiments with concrete were completed in the second half of the nineteenth century, when many intellectual models of the material's mechanism were developed, but it was one by François Hennebique (1842–1921) that was eventually universally adopted to become the framework of our understanding of reinforced concrete's behaviour. Hennebique's bridges were simple arches that put the material in compression or employed elementary reinforced beams, and it was through a manipulation of a concrete arch bridge that a key refinement in the use of reinforced concrete came to be made by one of his disciples, Eugène Freyssinet (1879–1962).

In 1910 Freyssinet, a graduate of the Ecoles des Ponts et Chaussées, was a junior engineer in the highways department near Moulins in central France. Three failing suspension bridges over the Allier River were to be replaced with stone arches, but initial budgets indicated that only one could be covered. Freyssinet made an unsolicited bid to build all three in reinforced concrete for less than the budget, and was awarded the Veurdre bridges project.

His compression arches, a form of great antiquity, were low and slender. Concrete creeps and subsides under load, only gradually settling down and hardening to immobility. This effect had to be dealt with in order to achieve such low arches, or the rings would have sagged. Freyssinet resolved the problem brilliantly. He left the tops of the arches open, separated by very strong jacks and, as the ribs settled, he pushed them back up again. He was 50 by the time the bridges were finished and it was only then that it occurred to him that stressing concrete before service loads were applied could be of much wider application. He rapidly developed a method in which cables were laced through ducts and pulled tight or rods were stretched and then concrete cast around them to make monolithic units that would not crack.

The cheapness of the material, its longevity in warmer climates and, in the early days, its requirement for intensive skilled labour to make formwork, set bars, and carefully mix and compact pours, made the uptake of reinforced concrete in Italy, Spain and Mexico particularly rapid. Because concrete is poured it also has the potential to be moulded into plastic form. The modern treatment of a bridge as a designed object capable of expression, through plastic, sculptural effects and the externalization of inner mechanisms (supposed force patterns), came into being as part of this development phase of reinforced-concrete structures in Latin countries, although Switzerland provided an additional epicentre of activity.

The pioneer was Eduardo Torroja (1899–1961), born in Madrid. As soon as he was qualified, in 1923, he began working for himself and almost immediately adapted Freyssinet's innovations in prestressing to his own bridge-building. He worked intuitively and drew on a practical inheritance from his father's contracting firm, claiming that design calculations were only necessary to prove his proposals, an approach still used by modern designers, who will 'put a design through the code' only when complete in its major parts. He was innovative in unexpected ways. His projects reveal their construction technique and combine a self-conscious concern for appearance with an unselfconscious eye for proportion.

In Italy Pier Luigi Nervi (1891–1979) transformed the use of reinforced concrete in building structures. He maintained that the production of

figure 13:
The German 'pocket battleship' *Admiral Scheer*, 1940, was a lightweight raiding warship. Hull weights for this kind of ship were greatly reduced following developments in electric arc welding. The resulting ships were bigger than the Treaty of Versailles had intended.

figure 14:
A photograph showing the US ship *Schenectady* split down the middle. The brittle failure of large fully welded steel structures – especially those of sectionalized construction that had been subjected to frequent load reversals – was not uncommon.

form is identical for technicians and artists; calculations are simply the method chosen to bring forth form – and there are many methods of calculation to chose from.

Ricardo Morandi (1902–1990), from Rome, displayed a very modern sensibility with reinforced concrete during his very long career. His work developed a muscular and idiosyncratic style and he appears to have integrated the ascendant architectural style of brutalism into his engineering. In the late 1950s he won a competition for the Lake Maracibo Crossing in Venezuela, his prestressed-concrete scheme beating 11 other rivals, who all proposed steel structures. On the huge viaduct at Maracibo the material is directly expressed, the design is not over-mannered and there is a fascination with the way in which details can be made to work, not just with how they look.

The later twentieth century: new tools and sensibilities
The groundwork for our current preoccupations with aesthetics was laid before the Second World War. This conflict and the subsequent regeneration also spurred on new ways of building bridges in subsequent decades, and the advances that define our age and give it conceptual shape could be viewed as a product of the Cold War arms race.

Robert Maillart epitomizes the technically competent, rationalist engineer acutely conscious of the aesthetic potential of bridges. He was thoroughly absorbed in his engineering work, which was based on a good theoretical grounding from studying under the academic bridge consultants ascendant in Switzerland at the turn of the century, followed by practical experience working for Hennebique, then pioneering the new structural material, reinforced concrete. A life of continuous experimentation gave rise to a whole series of innovations in the use of reinforced concrete and construction methods. His independent spirit asserted itself with the foundation of his own firm in 1902, in Zurich, and his departure a decade later to build in Russia, from where he returned impoverished after the October Revolution.

None of Maillart's bridges is very large. They fit into their often spectacular locations with an apparent effortlessness (for example the Tavanasa Bridge, fig. 11), and there is an inevitability and elegance in their form. The spareness stems from their economic background, often the products of competitive bids made to small Swiss cantons with limited resources. The Calvinist environment made an elimination of the superfluous almost innate. Careful observation in the field, studies of cracking in plaster models, and the deliberate expression of the diagrams he had been taught to use for analysis (fig. 12), all

showed Maillart where redundant material could be removed. This paring down process led to a distinctive bridge form that has become a classic, helped by extremely photogenic settings.

He exploited the good rock foundations in the mountain gorges to hold in the spreading forces produced by his preferred structural system, the shallow three-pinned arch. Reinforced concrete, his favourite material, suited the remote sites. Whereas schemes in steel required large prefabricated elements to be taken up twisting roads, concrete could be cast in free forms from material found nearby, and using local stone led to a colour match with the surrounding rocks. Maintenance requirements were negligible, which suited the relatively remote locations. Maillart did not overtly seek sculptural expression but moved intuitively within a technical milieu to find his forms. His was the first portfolio of an engineer's work to be exhibited at the New York Museum of Modern Art, in 1957.

At the same time as the sensibilities demonstrated by Robert Maillart were becoming recognized, large-scale changes in techniques and economic scale were affecting bridge design. The conclusion of the Second World War brought about the widespread destruction of bridges throughout Europe, particularly over the great rivers flowing north through Germany. Meanwhile the economic expansion of the

United States after its victory led to new and ever bigger crossings in North America.

Two strands of development appeared, separated by the Atlantic. In Europe the Marshall Plan (European Recovery Programme) required replacement bridges as soon as possible. New links were called for to open up Italy and the Adriatic as a bulwark to the east. Medium-span bridges and high viaducts were required that would be simple to erect and use a minimum of scarce resources.

The war years had seen the transformation of large-scale steel fabrication methods and these were quickly taken up in the reconstruction. The advances in technology originated in German efforts to circumvent the Treaty of Versailles of 1919, which had limited their capital ships to a weight of 11,200 tonnes (11,000 tons). Steel hulls were originally made by overlapping steel plates like scales and then riveting them together, but if plates could be butted end on end the extra weight of the lap could be eliminated. This was achieved by improving electric arc welding techniques and the 'pocket battleship' was born (fig. 13). All steel fabrications are now made this way.

The vaunting of this ingenious application contained some misleading propaganda, however. Welds tend to be brittle and the huge heat generated by the process locks in distortions and residual stresses. The Allies mimicked the method for their Liberty ships bringing North American supplies to the European war effort and extended it by introducing sectionalized construction: large repetitive sub-assemblies brought together for speed of fabrication. The ships therefore sometimes split in heavy seas (fig. 14), but despite this the margins prevailing in the Battle of the Atlantic made the 'teething problem' acceptable. After the war the remainder sale of shipping made Greek shipowners into tycoons and ensured the universal acceptance and continued development of the fully welded system.

Steel fabrication technology was readily transferred to bridge construction. The Cologne-Deutz bridge of 1946 spans 136m (446ft) across the Rhine and is formed from welded box sections built straight onto the old piers of a suspension bridge destroyed in the bombing (fig.15). Only a decade later the Sava Road Bridge in Belgrade was built in the same way to the huge span of 261m (856ft).

Further north along the German rivers the scale of crossing increased to an even greater magnitude. A very old form, the cable-stayed bridge, was reviewed and brought up to date, initially by the German Franz Dischinger (1887–1953). Downstream from Cologne in Dusseldorf during the early 1950s, Fritz Leonhardt adopted the form for a family of three bridges each spanning upwards of 280m (919ft).

The method of achieving such spans is simple. The effectiveness of a concrete or welded-steel girder can be extended by adding a superstructure of stays and the deck is supported on an array of cables carried over one or more towers. The system forms a satisfactory intermediate solution between a box girder and a suspension bridge – the longest form of bridge possible. The number of elements in a stayed bridge is low compared with other types, an important economy as cost resides in the details. The method of construction is relatively simple, with the two sides being built out towards one another sequentially.

The appearance of the bridges depends very much on overall proportioning and on the positioning of the stay groups. Harp and fan arrangements, for instance, have been shown to be of similar efficiency. The early German examples used a small number of elements to achieve an elegance that makes the lower Rhine crossings so distinctive, making them symbols of the *Wirtschaftswunder*, the successful post-war reconstruction. Bigger and bigger recent examples have much larger numbers of stays reduced to ethereal arrays of wire gathering onto monumental trestle towers.

The North American experience has been quite different, however. On the approach to the Second World War two particular influences

figure 15:
The Cologne-Deutz Bridge, 1948. Using money from the Marshall Plan, wartime advances in welding and sectional construction were put to immediate peace-time use.

figure 16:
The George Washington Bridge, New York, 1931, by Othmar Ammann. This truly modern bridge project was developed from sophisticated structural theory and with an awareness of the principles of industrial design.

made themselves felt. Firstly, industrial design was becoming a recognized discipline and bridge design formed part of this new awareness. By 1930 an earlier attitude of almost innocent pragmatism among engineers was being replaced by a quest for refinement, comprising components of functionalism, truth to structural action and, at the same time, a new interest in streamlining – the imposition of smooth, sometimes false forms onto objects.

Secondly, economics affected prevalent sensibilities. The immediate effect of the Wall Street Crash of 1929 can be seen in the George Washington suspension bridge over the Hudson (fig. 16), a definitive structure of its time. With a main span of 1067m (3501ft) it was nearly twice as long as any predecessor. The designer, Othmar Ammann (1879–1965), a Swiss émigré, used a sophisticated 'deflection theory' in his calculations to show that a relatively thin deck girder would be adequate rather than the deep stiffening trusses then the norm for large-scale suspension bridges. The long-lost lightness and grace of early suspension structures was therefore recovered. The design was originally to have monumental masonry piers formed from stone cladding on a grillage of steel, but a popular enthusiasm for the unadorned structure coincided with the cash shortfall of the Depression and the bridge's exposed steelwork became the new aesthetic standard.

The next major bridge in America, Joseph Strauss's Golden Gate Bridge in San Francisco (1933–7; fig. 17), could be seen as the structure that shifts bridge design into its modern phase. The project is iconic: its siting between rolling hills, the climatic conditions and prevailing mists, and its orientation across the line of the setting sun, together with the bridge's unusual height (to allow the passage of naval ships) and the dark ochre of the red-lead primer on an all-metal structure, combine to yield photographs that have made the structure the universal example of early modern bridge design.

The Golden Gate Bridge answered a direct economic need to open up Marin County to the city. It is technically extreme, both in its superstructure and in its foundations. The eastern tower is founded on unstable schists within the tidal flow and uncertainty persisted throughout the construction phase that a foundation could be made at all.

The designer displayed the intensity of will typical of bridge engineers. He had to lobby for 15 years before the project began but his early designs (fig. 18), hindered by his received knowledge as a bascule-bridge designer (fig. 19), benefited from the enforced wait. The project's architect Irving Foster Morrow (born 1884) worked hard to resolve and simplify the scheme, his efforts culminating in the utter

clarity of the built design. The spare ornament to the towers is instantly recognizable.

Paradigms of modern bridge types now began to be set up and aesthetic response became a recognized part of the engineer's remit. Alongside these growing sensibilities, material advances and technical innovations, other definitive components of today's practice took their place. New design tools have made it possible to describe and analyse any form accurately, to determine behaviour under different conditions and over time exactly, so that we are now free to test anything we can imagine before we build it.

The finite element method of solving partial differential equations has been known for some time; inklings of it appear in the work of Gottfried Leibniz in the 1660s. Differential equations deal with rates of change, whether in terms of flowing water, a planet's pull on a satellite or the strains in a solid object under load. They are mathematical models of physical situations and, manipulated on paper, they can show engineers how the systems they propose to set up will behave. Unfortunately most of these equations do not have ready solutions, although some can be elegantly resolved by hand calculation. Classically trained French engineers in the seventeenth century, intoxicated by the forms within the mathematics rather than on the

ground, would design their structures to coincide with a graceful suite of equations.

This process of developing a design around the requirements of a meta-system, making the form fit a way of doing things, pervades engineering. Most bridge forms can be seen as reflections of ways of building or as attempts by engineers to marshall force systems into understandable patterns.

As the name implies, finite element analysis involves imagining a continuous structure as a set of pieces joined by imaginary boundaries. In using this method, the mathematician can avoid the complexities of governing differential equations by defining the edges algebraically, namely whether they stay together or maintain some other relationship such as cracking apart. The result is a huge array of very simple equations that can be laboriously solved. Now, however, any differential equation for which solutions exist can be worked out on computer and this limitation on form has gone.

Pure mathematicians brought these numerical methods into the twentieth century and engineers began tentatively to use them. The second Aswan dam of 1910 was the first major practical application. Very large civil engineering projects could tolerate the staffing levels needed for the time-consuming process, but the approaching revolution

had to wait for a further two components: a means of dealing with the drudgery of the arithmetic involved and an industry with a pressing need for the technique and money to pay for its development.

The meteoric development of the electronic computer out of the code-breaking machines of the Second World War is well documented. It has delivered the capacity of a whole army of 'stressmen', men and women employed in the 1920s and 1930s to carry out repetitive sums on adding machines, to the individual engineer's desktop. The resource problem has evaporated.

The aerospace industry provided the budget and the pressing need. By the end of the Second World War the huge bombing campaigns had put in place an industrial base capable of producing large long-distance airframes. The Allied effort had opened up a transatlantic axis that initiated international air travel on an ever-expanding scale. Even more elaborate machines were needed for the arms race and space exploration. The structures had to be light and efficient, operating in environments susceptible to dynamic effects, extreme temperature and metal fatigue (fig. 20). The wealth of the aerospace industry allowed purer research than usual at a critical time, and several workers developed similar techniques for the analysis of airframes independent of one another.

Mathematicians set out the shape of the finite element method. The American researcher Richard Courant (1888–1972) is credited with a definitive exposition in 1943, but the way in which engineers now organize their understanding of the approach stems from a paper by four Boeing engineers, Turner, Clough, Martin and Topp, published in 1956. Developments followed piecemeal, as pressing problems required practical solutions. It was gradually understood that non-linear behaviour (uneven and sudden changes), instabilities such as buckling and dynamic effects such as vibration were all amenable to the method. The mathematical underpinnings of the subject caught up in the 1970s, as the precocious techniques of the engineers were regularized.

Textbooks had therefore been established and the method was in widespread use when the key revelation appeared in 1972. The Czech mathematician Ivo Babuska (born 1926) showed that the method was truly universal in application, and 'packages' of general purpose programmes began to appear. A modern computer can now readily determine the conditions within a structure comprising several different materials, joined together in any chosen way, and then go on to chart its behaviour over time, including the occurrence of instabilities, cracking and creep effects.

17 | 18

figure 17:
The Golden Gate Bridge, San Francisco, 1933–7, by Joseph Strauss and Irving Morrow: one of the most famous and photogenic of the world's big crossings.

figure 18:
An early scheme design for the Golden Gate bridge, 1924, by Joseph Strauss. The design was subsequently simplified and clarified, demonstrating how Strauss ultimately freed himself from the shackles of received knowledge.

The fact that airframe structures are so tightly determined by the constraints of flight is particularly significant. The structures cannot be idealized to respond to the rationale of the forces generated because the load-bearing elements of a fuselage are confined between the streamlined hull and payload space. The defining criteria of the airframe are external to the structural logic, and this can also be the case in bridge design. As the analytical methods of the aircraft engineer have become inexpensive enough to apply to bridges, so the bonds of structural 'sense' are loosening.

1878–2000: A catalogue of disasters
One model of modern bridge-engineering proposes that it is formed from an accretion of different understandings and preconceptions. Opposing this is the idea of practice as a series of paradigm shifts within an ever-increasing scale and level of sophistication in terms of project and procurement.

A third view identifies today's bridge engineering as a field undergoing cycles of change, including reaction, exploration, complacency and setback. Bridges may get bigger but the pattern of practice remains constant. Running through the broad weave of progress is, of course, the dark thread of disaster. Collapses occur, lives are lost and careers ruined. Each advance in knowledge is then tempered by the inevitable reactions and a return to more conservative modes.

The response to the Ashtalabula Bridge failure of 1876, a raft of legislation, has already been mentioned (see p.25), but the first Tay Bridge disaster in Scotland, three years later, better reflects the cyclical nature of advance in applied technology. In the preceding 20 years profound innovations had been pioneered. The uncertainties of change forced meticulous attention to detail and care in specifying materials and controlling the quality of construction, so that any problems had now been solved. The railway boom refocused competition away from work on the boundaries of technology towards consolidation and the economy of proven systems. It is a misapprehension to assume that periods of intense competition lead to technical innovation. Avoiding even limited damage, which could cause ruin, becomes paramount. Complacency also sets in and designs are appropriated by a second generation of designers less focused on the single project and more on volume production. Design codes prove highly useful in these conditions, and restricted practice in turn perpetuates the conservative mentality.

The Tay collapse exhibits a recurrent feature of major disasters: a confluence of mistakes and contributory factors. In the second half of the nineteenth century two rival railway companies linked Scotland and England. A series of takeovers gave the upper hand to the Caledonian Railway and it became clear that if the North British Railway was to survive then the firths of the Tay and the Forth would have to be crossed. The Tay was the shallower estuary and the engineer Thomas Bouch (1822–90) proposed the proven system of wrought-iron trusses on trestles but with many more spans than usual. Construction went very badly and during the six years it took to build the bridge 20 workers were killed. Foundations proved difficult to set up but, even more importantly, the main trusses – prefabricated, floated out and lifted hydraulically in the manner pioneered by Stephenson and then Brunel 20 years earlier – kept getting stuck. The bridge was finally opened in May 1878 and a year later Queen Victoria crossed in the Royal Train.

In the first 18 months of service the North British Railway's cash flow had not recovered sufficiently to implement a competent

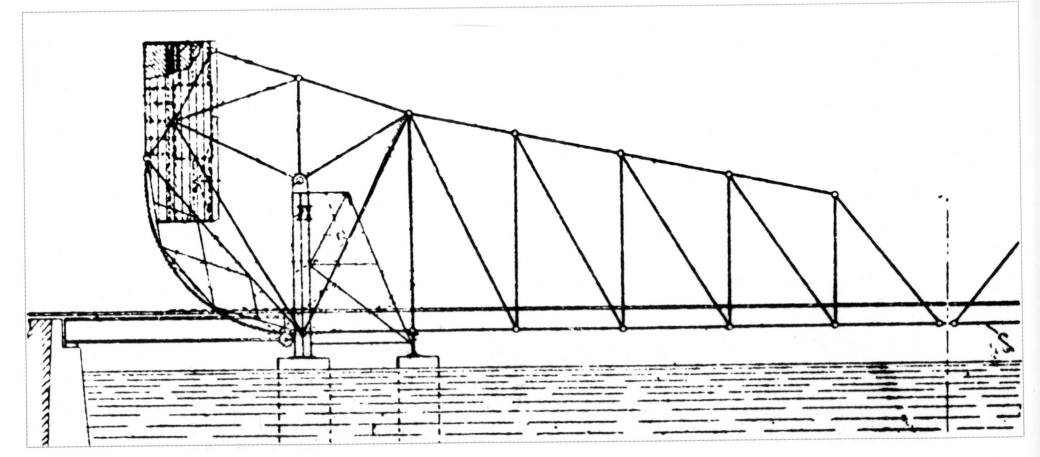

maintenance and inspection regime. At 7pm on 28 December 1879, two hours into a large storm blowing in from Scandinavia, a northbound train slowly tried to cross the bridge. As soon as it entered the middle section of the bridge the entire structure collapsed, taking with it 75 passengers and crew (fig. 21).

Once it lay in pieces the as-built condition of the structure could not be determined, but much of the material recovered was found to be defective. Many of the principle cast-iron elements had large blow holes that had been finished with a primitive adhesive filler known as Beaumont's egg. It could be argued that the danger of cavities, which act as serious stress concentrators in metal components, was not established until elastic theory was developed in the 1890s, but the efforts made to conceal these defects at the time certainly indicate some recognition of the potential problem.

Although the quality of the workmanship had been compromised by asking too much of the contractor and by a general under-capitalization of the project, it should be recognized that the design itself contained glaring faults. The tracks curved out over the estuary on paired girders. In the middle of the crossing the girders were set up

higher so that the train passed within them and the air draft for shipping below was correspondingly increased. Although it now seems unbelievable, no connection capable of resisting lateral forces was made between this centre section and the side spans. Most important of all, the assessment of wind load on the bridge was completely inadequate.

The power of the wind inevitably varies with the weather. Storm conditions (Beaufort scale 8) are identified with winds blowing at about 80kph (50mph). These press on a flat plate with a pressure of about 50kg per square metre (10lb per square foot), half the load that a house floor might bear. Surroundings affect conditions, however, since buildings and trees chop the wind into gusts and long, empty reaches allow the wind to build up power. In Britain the influence of the Atlantic brings prevailing winds from the south-west, but the next most common winds are from the opposite direction and these north-easterlies, which come off the North Sea, are of extreme ferocity.

Bouch relied on wind load tables prepared by John Smeaton (1724–92), the man identified as the founder of British civil engineering. This decision fatally compounded Bouch's complacency because the

tables made no allowance for the critical aspect of wind and structure interaction. The energy of wind forms a spectrum of frequencies to which structures respond differently so that, in any given wind, a stone structure will draw in less energy than a metal one. If a design assumes a static pressure condition, an even push against the side of an object rather than a transfer of energy, then the 60kg per square metre (12lb per square foot) that Smeaton proposed for the sides of his masonry structures is a safe calculation. However, Bouch applied Smeaton's figure to open wrought-iron trusses.

Allowing for wind and structure interaction, the equivalent static wind pressure on the Tay Bridge that December night was probably about 250kg per square metre (50lb per square foot), the equivalent of an office floor load and five times as large as the figure that Bouch had provided for in the bridge. Traditionally, overall safety factors are set at about three times the anticipated figure, so this allowance was completely overwhelmed. Following the finds of the board of enquiry, wind assessments began to be treated with the utmost respect.

The Forth Railway Bridge (1882–9; fig. 22) is, by contrast, the most beautiful embodiment of conservatism and caution. It represents a

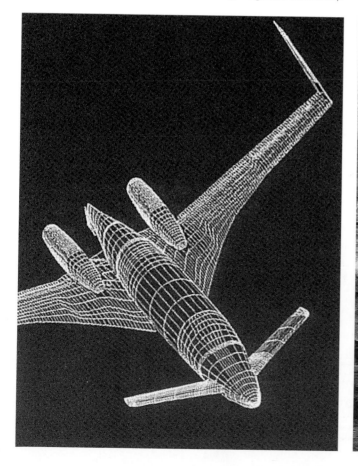

figure 19:
Strauss's 1915 patent drawing of a bascule opening bridge.

figure 20:
An aeronautical structure reduced to a wireframe model for computerized stress analysis. Commercial aviation, space exploration and the Cold War required ever more efficient structures. This procedure made the thinnest assemblies accessible.

figure 21:
The aftermath of the Tay Bridge disaster on 28 December 1879. A crossing train, the central span and the supporting trestles simply blew down in the high winds. Design inadequacies were compounded by poor details and bad construction practice.

reaction to the Tay disaster and epitomizes the stolid and sound aspects of the Victorian engineering establishment. Preliminary works had begun on Bouch's suspension bridge design over the Forth when further down the line the Tay Bridge failed. Bouch was brutally discredited by the board of enquiry and his new project was abandoned. The work then passed to John Fowler (1817–98) and Benjamin Baker (1840–1907), whose alternative design was innovative in Britain, since they proposed the use of steel throughout. James Eads had proved the material's use in the St Louis bridge eight years earlier, but a change in British legislation was required before the Clydeside ship-building expertise in steel fabrication could be transferred across Scotland.

Fowler and Baker developed and reviewed three further suspension-bridge options, but they were dissatisfied with the limited rigidity they could attain and were aware that there is little limit to the stiffness that can be achieved with continuous trussed girders. A cantilever bridge structure had just been used for the first time in a railway bridge in Germany and they decided to adopt this configuration, making their cantilevers very deep indeed. The final design weighed 50,000 tonnes (49,200 tons) and has been estimated to be over-structured by a factor of five.

For an assessment of wind loading the best expedient was adopted, by using site measurement. To do this, a small window from the north-east elevation of a cottage on one of the nearby islands was removed. The glass had survived for many years and was now loaded with sand until it cracked. The exercise was used to calculate the pressure sustained by the new bridge. As evidence of its success, the structure was refurbished for another century of use at the turn of the millennium.

The first Quebec Bridge (1904–7), over the St Lawrence River in Quebec City, was a cantilever design based closely on the arrangement of the Forth railway bridge. Its engineer, Theodore Cooper (1839–1919), had been Eads's diligent if somewhat panicky chief engineer at St Louis. As with the Forth Bridge the construction required the two sides to be built out towards each other. Early in August 1907, with each side about two-thirds completed, someone noticed that the plates at the bottom of the main struts were beginning to buckle out of shape. Cooper, by now old and infirm, visited rarely and supervised the work from New York. His order to the contractor to stop work arrived too late and, on 29 August 1907, the north side of the growing structure collapsed, taking 85 men with it. Only 11 survived, making this the worst construction accident of the twentieth century.

The wreckage lay in an almost neat pattern along the intended centre line (fig. 23). Investigators found the material to be of good quality and the workmanship excellent, but there were several apparently spontaneous failures in the main members and the enquiry concluded that the design would never have worked (fig. 24). Cost-cutting pressures had, however, contributed to the design's failure. The span of the original design had been extended by 15 per cent to bring the supports into shallower water, where foundations would be much cheaper, making it the longest span in the world.

After the extreme effort of completing a major design, revisions can often be undertaken in a less considered manner. The bottom struts of the Quebec Bridge had a more pronounced curve on elevation than the Forth Bridge, aggravating the additional forces that curved members introduce. The rectangular build-up of the main struts therefore proved critical. Fowler and Baker used standard North American design practice but extrapolated to a much larger scale,

adjusting their design methods to suit their chosen form. In contrast, the over-structured Forth Bridge had cellular tubular struts (like bamboo), an ideal section, adopted on the basis of observation and a cultivated intuition of structural form.

Structural elements loaded in compression, props and struts, have a theoretical capacity (strength) beyond which they will buckle. This strength is significantly reduced if the member is already slightly bent or contains plates that are not quite flat. The hot-working required to make metal structures – cutting, riveting and welding – leaves locked-in stresses. If the pattern is uneven, a hidden propensity to distort is built in. Rectangular plate layouts have proved particularly sensitive to these problems. The flat plates in the Quebec Bridge buckled locally and the struts followed this instability. Cooper's undoubted competence and experience were caught out by commercial pressure and the absence of a clear-headed theoretical overview. He died a broken man five years after the disaster. The replacement design used over two and a half times as much steel as the first scheme and reverted to straight bottom chords. A massive centre span was prefabricated and floated out to be lifted into place, minimizing stresses at the construction stage, but this too fell, when a stirrup casting failed during jacking, and another 11 men were lost.

Of the major bridge disasters, most have occurred during construction. A process of 'covert innovation', slight extensions of established practice that prove crucial, is evident in a series of failures of steel box girders in the late 1960s and the 1970s. In this boom period they seemed to offer economic, reliable and readily buildable bridges suited to medium-span crossings. Despite these factors the familiar conditions of economic pressure, established design practice, extensive precedent and complacency took their toll.

Typically, a steel through-box girder bridge comprises a series of prefabricated units and is erected by cantilevering them out towards each other from the two sides. Units are passed out along the cantilever or lifted up from below and welded into place. Early in June 1970 the bridge under construction at Milford Haven, Wales, collapsed killing four people. Three months later a very similar design over the Yarra at Melbourne, Australia, spontaneously split across its centre joint under very little load and 35 people were lost. The apparent simplicity in the analysis of both of these forms had proved deceptive.

At Milford Haven a plate cross-frame had collapsed and welded joints had then 'unzipped'; in Australia the welding had failed. The projects were not seen as departing from previous practice but they had in fact accrued subtle but significant differences.

In fuselages and ships' hulls the tubes are short compared with their girth. In bridges the ratios of length to width are much higher. When an unsymmetrical tube, like a bridge deck, is twisted its longitudinal fibres distort and it warps. The sensitivity of bridge forms to this phenomenon was underestimated at the time of the Milford Haven and Melbourne disasters. Compared with the general design it is inordinately difficult to assess. Cross-sections are forced out of plane and this effect, coupled with the high local forces incoming at bearing points, cause the structure to buckle over its supports.

Welding difficulties appeared as more and more site welds were undertaken. Good welds require very accurate fit-up between plates. Metal arc welding acts by striking an electrical charge between the welding torch and the plates to be joined. The intense heat around the continuous spark melds the metal together. Oxygen must not get into the molten metal or the result is a brittle zone, but methods have been

figure 22:
The Forth Railway Bridge by John Fowler and Benjamin Baker, 1882–9: a combination of precocious Victorian engineering, Clydeside shipbuilding expertise and the improved economic circumstances after the railway boom was over.

figure 23:
The north span of the Quebec Bridge, shortly after it collapsed on 29 August, 1907. Local buckling in the partially completed structure was first observed two months earlier. Instructions to stop work were fatally overruled.

developed to avoid this problem. One technique feeds inert material onto the surface of the molten pool to cloak it from the air, another more recent one blows gases over the entire work area to shield it. While a good match and draft-free conditions can be readily attained in a shipyard, they are much more difficult to achieve on an exposed bridge. There is currently no technical way of predicting distortion and residual stresses in welded assemblies.

The spate of major suspension bridges built just before the Second World War represents the work of a handful of designers. The 'heavy-weights', and Othmar Ammann in particular, poured material and stiffness into their structures. Ammann, an apprentice of Gustav Lindenthal (1850–1935), sought classical proportions that were not incompatible with mass. His towers are simple, with single top spandrels, and the longitudinal girders are enclosed trusses or single plate sections.

Meanwhile, a younger breed began to seek economy even in the largest suspension bridges. Ammann's rival, David Steinman (1887–1960), another protégé of Lindenthal, pursued slenderness and grace, decorating his suspension bridges, particularly the towers, with pierced patterns. A third designer, Leon Moissieff, responded to the influence of both men and sought to compete through innovation. In 1940 he built a bridge of fine proportions at Tacoma Narrows in Washington State, employing a very high deck with solid side beams and a stripped-down elegance. Its structural weight was very low in comparison with similar spans.

As soon as the bridge went into service it exhibited a tendency to buck and shake even in moderate winds, earning the epithet of 'Galloping Gertie'; it even became a weekend destination for joy riders. Stays were added but to little effect. Wind-tunnel tests were then underway when, on 7 November 1940, in a steady, strong breeze, the structure took up a sequence of oscillation modes (fig. 25). Over several hours movements built up to an astounding writhing action until the deck was torn apart and collapsed into the river. Professor Burt Farquarson of the University of Washington had been responsible for checking the stability of the structure and was on site at the time. He was the last to leave and was deeply shocked when the structure did not weather the disturbance. There were no casualties apart from a dog abandoned in a car left on the bridge at the outset.

On its last day the bridge had run through the gamut of motions. The solid-sided deck was shallow relative to its width and such cross-sections act as crude but effective aerofoils. At a specific wind speed the deck starts to fly and therefore rises, but the stiffness of the system resists this motion and pulls it back. The bridge then flies in the other direction and is pulled back once again. This sequence falls into an oscillation and the bridge begins to gallop up and down, a motion that had been observed many times in various bridges before the Tacoma Narrows collapse.

The motion of the deck displaces the surrounding air, which calms the movement. Wind that has been disturbed passing near the ground or across broken country is turbulent and also tends to 'buffet out' any cyclical movement. Unfortunately the uninterrupted fetch up the Tacoma river, the funnelling effect of the Narrows and the height of the roadway above the water meant that the deck had been placed in an extremely even flow regime, which limited this aerodynamic damping.

A more serious effect stems from the aerofoil interacting with its own wake. A bluff object develops a low-pressure bubble immediately behind itself. This effect can be seen in the way dirt is sucked onto the

PERSPECTIVE VIEW OF BOTTOM CHORD SECTION OF QUEBEC BRIDGE, THAT FAILED

WEIGHT OF U.S.S. "BROOKLYN" 9215 TONS

CROSS-SECTION OF BOTTOM CHORD OF QUEBEC BRIDGE

CROSS-SECTION OF BOTTOM CHORD OF HELLGATE BRIDGE

CROSS-SECTION OF BOTTOM CHORD OF FORTH BRIDGE

SCIENTIFIC AMERICAN, N.Y.

back of a hatchback car or a cyclist. The bubble eventually implodes and then reforms, giving a light twist to the object in front of it. Once a bridge deck has been distorted, its angle of attack to the wind is changed and it flies once again. The twist is increased as it seeks to take off until it stalls and drops back through its own centre line, to take off downwards. Again a repetitive motion, known as 'stall flutter', is set up, which is damped by the passage of the wind. The Tacoma Narrows Bridge went through a period of this behaviour on its final day.

If the structural characteristics of the bridge allow it to vibrate naturally both vertically and by twisting at about the same frequency, then the motions described above couple together and the structure appears to hunt through the air like a fish swimming upstream. There is very little motion-damping and the structure begins to pick up energy on every cycle, however light the breeze, until no more can be absorbed and total instability occurs. When this process started at Tacoma Narrows, the end came quickly.

The phenomenon experienced at Tacoma Narrows was not new. A sketch from 1836 by an army major of Captain Samuel Brown's Brighton Chain Pier in the process of being blown apart accurately depicts one of the various forms of aerodynamic instability (fig. 26). Repeated additions were made to Telford's Menai Straits Bridge to steady the deck in the wind, while the Golden Gate Bridge exhibited a rippling mode and had to be stiffened laterally.

The fact that certain combinations of aerodynamic form and structural response will inevitable lead to failure, however strong or stiff the structure, was well known to aeronautical engineers. The problem had dominated early monoplane design at the end of the First World War, but despite this precedent there seems to have been a mental block among contemporary bridge engineers that prevented them from believing that the phenomenon could happen at such a large scale. The fact that the wind was not strong or buffeting at the moment of failure was remarked on repeatedly in reports. The loss of the Tacoma Narrows Bridge has since been interpreted as an example of how unforeseen effects can suddenly intrude on careful development, shifting design concerns into new areas.

The consequences of the collapse again demonstrate the processes that follow on from such setbacks. Moisseiff was then in the middle of a much larger plate-girder design for the crossing of the Mackinac Straits in Michigan. He was removed from the project and replaced by a triumvirate formed from Ammann, Steinman and a third engineer, Glenn B. Woodruff. After a contretemps Ammann withdrew and Steinman went on to produce the longest overall single bridge now in existence. The bridge was erected between 1954 and 1957 and founded on solid rock at great depth, with more massive foundations than had ever been used before. Icing was a regular occurrence and the superstructure was designed to withstand a huge load allowance of 16,000 kg per metre run (50 tons per foot run). This provision was ten times the amount that could reasonably be expected.

Ammann's last and greatest masterpiece, the Verrazano Narrows Bridge of 1959–64 in New York, was an even more massive structure for a comparatively benign environment (fig. 27). The simple tower profiles and clean lines disguise a huge mass and impose a strong sense of repose and stability.

In Europe the response to the Tacoma Narrows event was very different. Rather than add weight and stiffness to bridges to preclude all aerodynamic behaviour, several engineers sought to understand

24 | 25 | 26

figure 24:
Illustration from *Scientific American* (1907) explaining why the Quebec bridge design would never have worked.

figure 25:
The Tacoma Narrows Bridge in the penultimate phase of its collapse, 7 November 1940. A vertical 'galloping' motion coupled with a twisting 'stall flutter' created the most dangerous form of aerodynamic instability: 'classical flutter'.

figure 26:
An eyewitness sketch illustrating the collapse of Samuel Brown's Brighton Chain Pier (1836).

the problem and manipulate the effects that had presented themselves. Their theoretical insights produced a solution. The engineer Fritz Leonhardt proposed an aerofoil deck section that would stay steady in an airstream with undisturbed air flows passing around it. He put forward this idea for the competition for the Tagus crossing in Lisbon (1959), but it was eventually won by a deep truss bridge designed by an American consortium including Steinman.

Some years earlier, in England, a partnership of consultants Mott Hay and Anderson and the firm Freeman Fox and Partners had been appointed to design a crossing for the Severn. They had worked previously on the Forth Road Bridge, a trussed-deck suspension bridge with some innovation in the tubular towers (fig. 28). In 1960 work had commenced on building a similar design across the Severn when research appeared by the Norwegian Arne Selberg which conclusively showed that long bridges act rather like aircraft wings or sails. Although they flap violently if not controlled, they become solid and steady when properly arranged in an airflow. An engineer at Freeman Fox and Partners, Gilbert Roberts, took on Leonhardt's idea and developed it through a series of wind tunnel

tests. A very light box girder profile, torsionally stiff, was worked up and implemented. Fuelled by the confidence of the times, several advances were therefore made at once, at remarkable speed.

The thin deck reduced the wind loads passed to the towers, which in turn became finer than usual. The Welsh architect Percy Thomas applied his classical training to the proportioning. Late on in the design the engineers became concerned and decided to make the main cables curve more than usual in order to improve their strength, further enhancing the appearance (fig. 29). Regrettably the scheme had overreached itself. A bridge should last a century, but within a decade serious inadequacies were showing in the Severn crossing, and traffic restrictions and remedial works followed.

The penalties of failure in this field are, inevitably, great. Both the scale of the forces mobilized and the fact that the work must be carried out in a social context makes bridge-building a political activity. Many protagonists have suffered broken careers, ill health and shattered dreams. As an extreme example, Herodotus (485–425BC) records that the Greek general Xerxes beheaded his engineers when a pontoon bridge over the Hellespont blew away (although this was

apparently not merely out of pique but also to propitiate the gods). Washington Roebling had to watch the Brooklyn Bridge being completed from a wheelchair in a loft room above the Hudson. He had been crippled by decompression sickness after a fire had confined him and some of his men for too long in a pressure caisson on the river bed.

The career of Claude Navier, a brilliant mathematician, never recovered when the Pont des Invalides was cancelled. The French consulting system had constructed an exact method of describing the design, assigning responsibilities and contractual arrangements. When difficulties arose the contractor, Desjardins, was able to show that the design had been followed to the letter and, as a result, liability was not spread evenly but fell completely on the designer. Navier was a representative of the elite corps of the Ecole des Ponts et Chaussées, the dominant power structure in engineering at that time, and its opposition seized on the incident and magnified it. The fact that the problem lay in the foundations and occurred at the end of the year meant that it could not be rectified until the following spring. This was, regrettably, represented as an inability to solve the problem and the bridge was rebuilt to a contractor's design. Navier continued at the

Ecole without promotion, and died in 1836 at the age of 42. His contemporaries noted his broken health and spirit.

Exactly one hundred years later, soon after construction began on the Golden Gate Bridge, Joseph Strauss suffered a breakdown. He was hounded by a press campaign questioning his title to the design and resigned within four months of the bridge's completion. He retired to write doggerel verse about his experiences and died two years later at the age of 68.

Contemporary Bridge Design

The cutting edge of bridge design now broadens in three directions. There are yet bigger crossings to make, which will be paid for by and promote economic growth, linking ever larger conurbations. The size of a bridge has a theoretical limit, however. Strength goes up as the square of its dimension, the cross-section of the supporting element, while self-weight goes up as the cube of the dimension, its volume. As structures get bigger they eventually come to a point where they collapse under their own weight. For steel the limit of the span would be 18km (11 miles), whereas for materials with

better strength to weight ratios correspondingly greater spans are possible.

There is still a long way to go before such limits are reached but new materials keep appearing, their application only restrained by concerns over longevity. Parafil ropes, astonishingly strong but also expensive and of doubtful longevity, have proved suitable as anchorage cables on deep-sea drilling rigs. Other plastics, with 'designer' structural properties invented for each application, are now within the economic range of practicality for beam and strut members. The initial proving period, in which new materials are used in conservative ways, is now well under way and completely new forms, stemming from a deeper understanding of the materials' possibilities, are to be expected shortly.

Besides further advances in material science another transformation is following: the revolution in the way load environments are assessed. This has always been an important area of risk management. All types of loading, not only wind, are now being considered as part of a spectrum of possibility to be statistically determined. If these realistic assessments are combined with models of structural reliability then

an overall safety rating can be identified and set economically. Structures will take on levels of attenuation and efficiency not seen previously.

Finally, the growing self-consciousness among bridge designers is producing a wide range of design approaches. A consensus as to what amounts to good bridge design has yet to emerge, but processes are at work. One tenet of modernism makes technology and its overt application a defining feature of new architecture. It has been a short step from engaging engineering as an essential part of architecture to appropriating engineering artifacts themselves as potential architectural products.

Architects and engineers may still be some way apart when it comes to agreeing upon what is admissable in bridge design, but the material is there for a critical superstructure to develop. This in turn can be used as a tool for understanding and evaluating the different approaches of the many talented designers involved in the field today. The following diverse range of examples are intended to reflect the breadth of ideas from which many new bridges, as beautiful and innovative as their predecessors, will spring.

figure 27:
From 1959 to 1981 the Verrazano Narrows Bridge, New York, boasted the world's longest span .

figure 28:
The Forth Road Bridge, 1958: a suspension bridge design relying on a deep stiffening girder beneath the road deck. The extreme attenuation of the towers was made possible by the introduction of thin-walled hollow steel struts.

figure 29:
The graceful lines of the first Severn Bridge, 1961. The conventional stiffened deck was replaced by an aerodynamically shaped welded plate box section.

30 Bridges: Case Studies

Helgeland Bridge; Alsten, Norway; Leonhardt, Andrä and Partner 1991
452m / 1,483 ft

Essential transport links can be difficult to develop in Norway's landscape of fjords and mountains but, to assist the process, wealth derived from shipping and North Sea oil has been reinvested in building the infrastructure necessary to make the country universally prosperous. The small, remote island of Alsten, 20km (12 miles) south of the Arctic Circle, has benefited from this investment and is now linked across the Leirfjord to the mainland road network.

As part of the North Sea oil exploration, advances have been made in handling wind and sea forces that have enabled this narrow two-lane highway bridge to become economically viable. Local weather conditions are extreme: gales coming off the North Sea gust around the islands and up the steep-walled fjords. Strong currents combine with storm waves, and temperatures regularly drop to 15°C (5°F). To add to this, complex rock formations along the coast reverberate with seismic activity from nearby in the Norwegian Sea.

The expansion of the oil industry into this harsh environment, where wind and waves are the dominant influences on structure, led to the rapid development of techniques to improve the assessment of likely wind loads, while, at the same time, reinforced-concrete construction was advanced in the largest rigs. When applied to bridge-building the new methods extend the margin at which seldom-used bridges become practicable.

In the past the wind tunnel has been an important tool for determining pressures. Wind effects have been dealt with by conjecturing the pressure to be resisted and then building a structure of the correct size. The chosen pressure depended not only on the wind but also on the shape of the structure. More recently, however, a subtler approach

that was initiated by the aeronautical industry, and appropriated by marine designers for deep-sea structures, has found its way into the wider fields of civil engineering.

Forms in an airstream interact with surrounding airflows, responding to these currents and also shaping them. In the process, energy is transferred from the air into the structure and back out again, in a movement that involves a spectrum of energies of different frequencies, depending on the nature of the wind. The structure, in turn, has a response spectrum: natural frequencies at which it vibrates and absorbs

energy. By combining the spectra of excitation and response, an assessment of the bridge's greatest anticipated acceleration can be gained and the movement kept within safe limits in the construction.

For the Helgeland Bridge these refinements were taken a significant step further. A weather station was set up on the bridge site and data compiled over three years. From measurements of the intensity and duration of storms, standard statistical models of the weather were adjusted to be specific to the site. A computer was then used to develop a projected future for the bridge, in which the effects of wind

were viewed over a certain period in terms of changes to the bridge's overall configuration. The idiosyncrasies of individual materials were also incorporated in the study. For example, steel bridges behave like springs but their concrete counterparts act more like malleable plastic masses.

These probability analyses reconfigure structural design as a type of risk assessment, identifying the likelihood of a certain situation. They also allow structures to be refined to a level of acceptability that may be set by economic or social forces. The upper criterion is

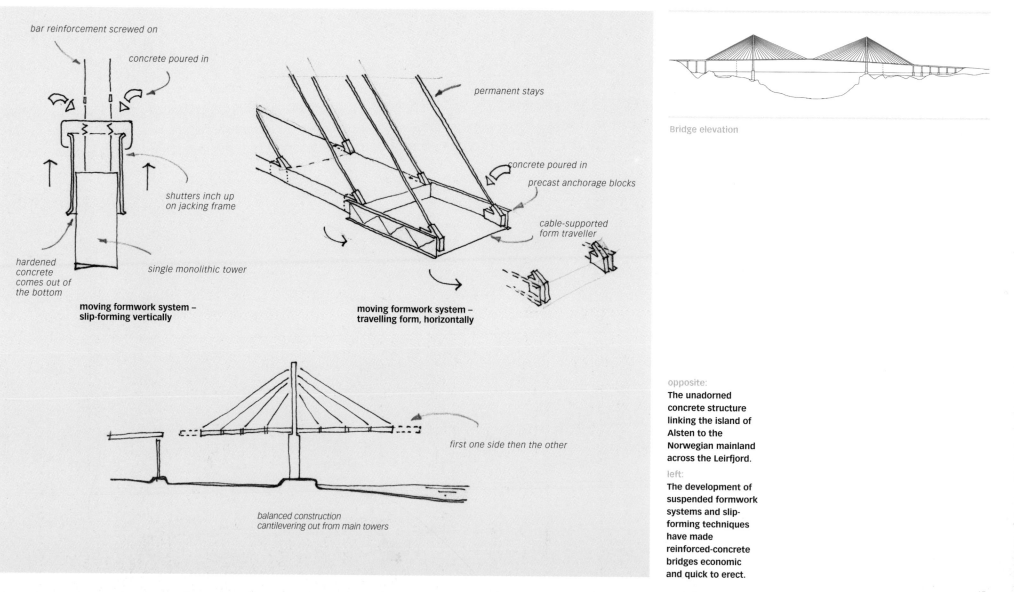

bar reinforcement screwed on

concrete poured in

hardened concrete comes out of the bottom

shutters inch up on jacking frame

single monolithic tower

moving formwork system – slip-forming vertically

permanent stays

concrete poured in

precast anchorage blocks

cable-supported form traveller

moving formwork system – travelling form, horizontally

first one side then the other

balanced construction cantilevering out from main towers

Bridge elevation

opposite:
The unadorned concrete structure linking the island of Alsten to the Norwegian mainland across the Leirfjord.

left:
The development of suspended formwork systems and slip-forming techniques have made reinforced-concrete bridges economic and quick to erect.

usually the '120-year storm'. This represents a probability level, a remote possibility at any one time, and the concept can be used to set much lower return periods during the bridge's life, co-ordinated with insurance levels.

The feasibility studies confirmed that wind loads would govern the design. Composite bridges combining steel and concrete were rejected, since a massive, monolithic concrete structure was shown to respond better to seismic loading and to be less prone to overturning. The scale of the crossing led to the consideration of cantilever and cable-stayed alternatives, of which a cable-stayed fan bridge with reinforced concrete towers and deck was found to be the least expensive and most durable option. The foundations on both sides were placed on hard rock, under 30m (98ft) of water on the north side, which avoided the expense of using lightweight concrete.

The surrounding area is sparsely populated and therefore the designer decided that the bridge should have a minimum maintenance regime. The deck was made extremely thin to reduce windage, and high-strength concrete (grade C65) was used throughout, the high-cement content improving durability. A continuous deck of concrete easily copes with the failure of a stay. Splayed towers are more resilient than H-frames in high winds, but diamond-shaped piers were chosen instead of A-frames on aesthetic grounds and because their bases fit neatly onto the narrow caissons. The bridge's tower section sizes are kept down by the tight layout of the upper anchorage blocks. The deck is restrained at each end only, not at the pier intersections, and therefore stabilizes the towers through the cable arrays rather than through the piers, allowing the uprights to be as thin as possible.

The construction of the bridge took two years, with work continuing throughout the subarctic winters. Very large barges (100x40m/328x131ft) were set up in the fjord as working platforms, with a concrete-making plant on board one vessel. The superstructure was constructed by the 'free cantilever' method, working outwards each way from the towers, and temporary auxiliary piers were added so that the balanced sides did not become too wide before receiving additional support. The towers were put up rapidly, cast as single monolithic pieces using slip-forms.

The very shallow deck was cast *in-situ*, as a giant sliding form. It weighed 115 tonnes (113 tons) and was temporarily supported on the permanent stays with precast anchor blocks to speed up the process of construction. The deck is not unduly stiff, which made the alignment corrections to bring the bridge sections together a straightforward process of stay adjustment. Unfortunately, however, sea spray penetrated the concrete despite the application of a plastic curing membrane (a layer painted on to envelope and protect the concrete as it hardens). As a result, plastic cracking, a defect occurring in the

first phase of concreting when the material shrinks too much before its strength is gained, was widespread.

The finished bridge has been festooned with meteorological equipment and transducers to record movement, so that the structure itself has become a sensitive instrument, measuring at first hand the forces that such an environment can generate. The data is being used not only to recalibrate the computer models made for the project, but also to inform the next generation of bridge designs.

opposite:
The balanced cantilevers extending outwards as construction proceeds in the summer of 1990. The work areas are protected by temporary awnings.

far left:
Arctic storm conditions encountered during winter work.

left:
Excellent rock foundations give the tower bases slim and elegant proportions.

Plashet Grove School Bridge; London, UK; Birds Portchmouth Russum / Techniker 2000
67m / 204ft

Plashet Grove secondary school in East London is separated into two halves by a curving, tree-lined avenue. The road is a busy bus route, the crossing made dangerous by a steadily increasing burden of traffic. To join the two halves of the school together and to avoid road danger, a footbridge that connects the two main accommodation blocks at first-floor level has been carefully inserted. Differing floor levels, end alignments and intervening trees are negotiated with a sinuous covered way.

The work of the architects Birds Portchmouth Russum is characterized by irreverent, highly contextual experiments with form. Several earlier bridge projects chart their enquiry. A sweeping footbridge links a car park to an old city wall; an urban motorway is given a sense of place with a suite of free-form bridges incorporating plastic support elements and applied reference systems. Recollections of an industrial archeology are evoked in such schemes, and the motif of a central break in a continuous form is developed.

The school bridge began with the reworking of a trussed tube design. A central break was added, which was then transformed into a gazebo. Cost constraints led to the reappraisal and steady simplification of the design, concluding in a simple pairing of rolled-steel beams, the largest standard size available, beneath a fabric canopy configured to suggest cloistered space.

These beams form the sides of the walkway and the standard balustrade height is achieved by adding an oversized rail along the top

flange. When a beam sags downwards the bottom edge is being stretched while the upper edge is being squeezed. The resulting compression makes the top flange want to flick sideways. In the original Plashet Grove scheme this effect was dealt with by connecting the side girders across their tops. In the final proposal the soft top required a different strategy. By connecting the two beams with a sufficiently rigid cross-frame on each side and below, the two upper flanges were adequately restrained. These U-frames rely on the deck being stiff. The necessary stiffening plates were integrated into the design of the cover supports to become intrinsic decoration.

It was particularly difficult to achieve the structure's twisting alignment. The beams were easily curved on plan, and were formed cold by being bent between rollers, but adjusting the vertical profile presented a greater problem. The assembly naturally has to be very stiff in the vertical plane, but cold rolling throws out the horizontal alignment. Initial proposals to overcome the problem involved jacking the bridge into position. The structure was to be assembled straight, then allowed to drop and finally forced into its finished alignment. However, the road closure requirement to complete this process was too long, and instead the beams were 'blacksmithed' into shape – heated, cooled and pulled – a craft process for which there are no written rules.

The structure is continuous from end to end to ensure that a smooth profile is maintained under all forms of partial loading. At over

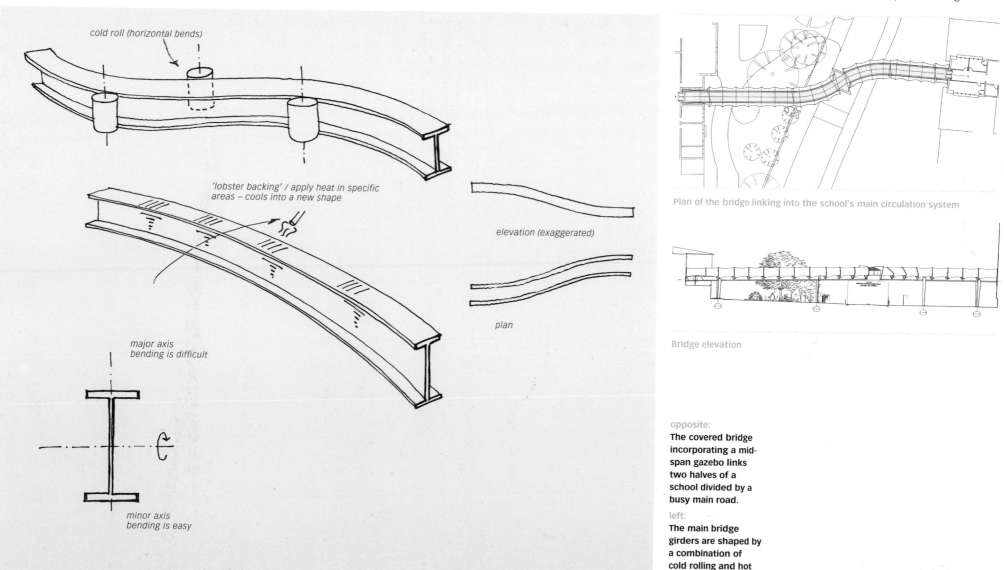

cold roll (horizontal bends)

'lobster backing' / apply heat in specific areas – cools into a new shape

elevation (exaggerated)

plan

major axis bending is difficult

minor axis bending is easy

Plan of the bridge linking into the school's main circulation system

Bridge elevation

opposite:
The covered bridge incorporating a mid-span gazebo links two halves of a school divided by a busy main road.

left:
The main bridge girders are shaped by a combination of cold rolling and hot blacksmithing.

49

60m (197ft) long, some form of expansion joint is essential. In the English climate a steel section of this size may expand and contract by up to 20mm (0.78in) in length, given the possible temperature variation of 40°C (104°F). The bridge is fixed back on the north side only, while an expansion joint at the south end is incorporated as a simple lap transferring only vertical and sideways forces.

The support piers are formalist elements developed by the architects with only occasional and slight interventions by the engineer. These simple plate structures are stiffened by curled edges and folds but the abstract shapes have a propensity to buckle, which can only be investigated by careful physical and computer-model study. Initial imperfections in the raw plate and local disturbing forces can set off a problem. Assessing the piers' sensitivity became key to the design effort.

The bridge's central gazebo is formed as a box of folded steel plate, welded between diaphragms attached to the walkway. The main beams pass beneath window seats, with plate-girder stiffeners carefully aligned to carry local forces safely around the disturbance. The welding of the intersecting plates is complex. Manual welding is required, a skill-sensitive process in which the operative drags a pool of molten metal into shape. It is difficult to ensure complete fusion between surfaces, to avoid contamination and to achieve an even cooling down without cracking. Critical welds need to be designed for testing. Butt welds are usually plumbed with ultrasound for tell-tale echoes that indicate hidden defects, but complex intersections of

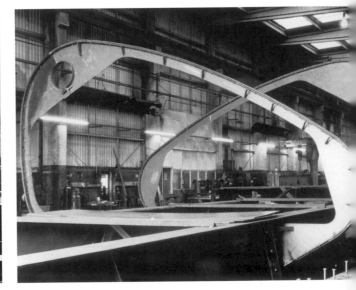

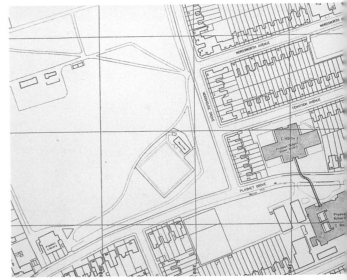

plate make interpreting the internal reflections fraught with difficulty. Defects are common and joints must be designed with an eye for the repair that may become necessary. The Plashet Grove bridge incorporates the necessary fabrication details thoughtfully. Starts and stops in a butt weld are discontinuities (weaknesses) and therefore cope holes are introduced, allowing welds to be passed from one welder to another across an intersecting plate. Run-out plates on edges allow the weld to be carried beyond the boundary to avoid

notching. The discarded ends, cut off when the welding is finished, can be etched to make a visual check of the work.

Architectural detailing on the bridge is intellectually robust. Beams are joined to piers with dramatic forelock pins. Rainwater is carried off the roof by stoma (tiny openings) in the membrane panels and runs into hoppers connected to the handrail, which operates as a downpipe.

The walkway covering comprises distorted paraboloids of Teflon-coated glass-fibre fabric stretched between shaped arch rings.

Bulkhead lights within the rings light the curved soffits from below. Panels are joined across steel-laced anchorages that are watertight without additional flaps or seals. With a translucency of some 18 per cent the canopy is adequately lit by day and glows at night.

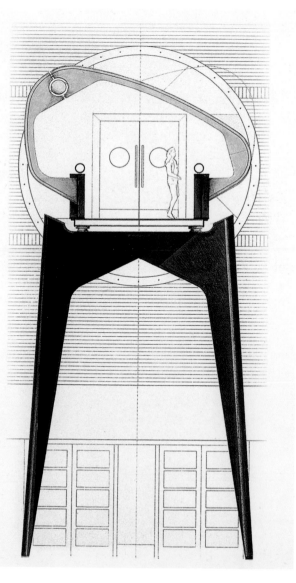

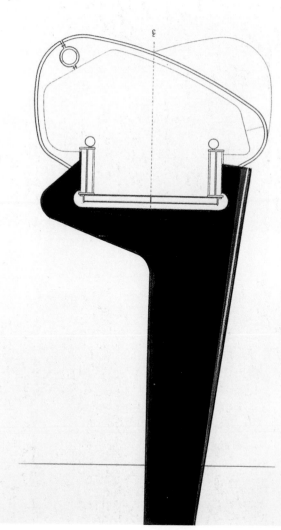

opposite, left:
The idiosyncratic roof form is made up of individually patterned Teflon-coated glass-fibre membrane panels.

opposite, top centre:
Each saddle-shaped enclosure becomes rigid when stressed. The weld marks along the seams bleach out in sunlight.

opposite, bottom centre:
The prefabricated steel skeleton was assembled in the playground and prepared for lifting into its final position.

opposite, top right:
The hoops supporting the roof are made from plate steel, water-cut on a computer-guided machine.

opposite, bottom right:
The comprehensive school is formed by the consolidation of earlier grammar and secondary schools. The bridge completes the link up.

far left:
An architectural study in pen and coloured pencil of the north pier and building junction.

left:
The piers are designed as sculptural forms in thick plate steel.

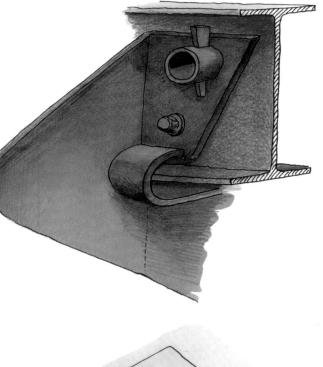

The cloister-like quality of the enclosure is emphasized by the translucency of the fabric and lighting provisions.

The central gazebo becomes part of the school's recreational space.

The primitive peg joint between the main girders and the piers.

Plates providing local stiffening are integrated into the column head design.

Greater New Orleans Bridge; New Orleans, Louisiana, USA; Modjeski and Masters 1984
480m / 1575 ft

By the time it reaches New Orleans the Mississippi river is flowing at a rate of more than a million cubic feet of water per second. Confined within man-made flood levees, the wide, slow river crosses a flat delta landscape of alluvial soils.

In response to these conditions road and rail crossings over the lower reaches of the river have evolved into a specific bridge form, now superseded elsewhere but still being developed in the Southern states. There is an unbroken line of descent leading from the Second Greater New Orleans Bridge, designated a 'through-truss cantilever bridge', back 150 years to the first cantilever railway bridges, of which the Forth Railway Bridge in Scotland is one of the earliest and largest examples.

The form of the Greater New Orleans Bridge reflects its surroundings. The river is an established transport route and spans have to be as wide as possible with high clearances for boats. Roadways must rise from the flat surroundings on long approach ramps. Foundations must go deep into the weak underlying alluviums and, for economy, must be concentrated into as few locations as possible.

Precedent pervades the design of the new bridge. The original cantilever bridges were developed to provide rigidity under the heavy loads of locomotives over ever-widening crossings. Each part was essential to the whole, so improvements in the consistency and quality of steel were key to the success of these bridges. A suspended middle span allowed the large thermal movements inevitable in metal structures to be readily accommodated on sliding joints. If such movements were confined, enormous bursting forces would occur.

The construction of such bridges was straightforward. The cantilever spans were built from the piers outwards, while the central span was prefabricated on a nearby bank, and then floated out and lifted into position.

Wide crossings generally occur on the lower sections of rivers where the silting adds to the problem of poor foundation conditions. Simple cantilever bridges sustain pier settlements without damaging changes to their internal force pattern. So the essential components of this bridge type were established: a minimum number of supports, a flat soffit providing a good air draft and a profiled upper chord responding to the pattern of the forces generated. A reliable method of construction was now available.

Development centred around keeping costs down. Material could be used more efficiently if the suspended span was made continuous with the anchor spans. Initially this was done by adding 'idle links'. Once a design had been resolved to suit the dead-load conditions of the finished structure, extra links were added which became operative only under live loads. The bridge worked a little more efficiently and the amount of material required could be reduced. It was not so easy for the designer to determine the pattern of forces within the structure, but confidence in handling methods of analyzing indeterminate structures improved as expertise was imported from Europe in the years before and after 1900. The simple control of thermal movements made possible by sliding joints was lost, but other means were found to allow for the required straining. The trestles, which were necessarily tall to let the riverboats pass underneath, could be made stiff enough to resist wind loads but soft enough to sway sufficiently for the bridge to expand and contract.

Once continuous girder configurations were proven, embellishments followed. Closure, the moment at which the two sides of a bridge are joined into a whole, is a critical stage of construction. During erection dimensions were continuously revised to bring the parts to their correct positions. Final adjustments were made by jacking (in the same way a car is jacked up) or by packing elements in hot coals or

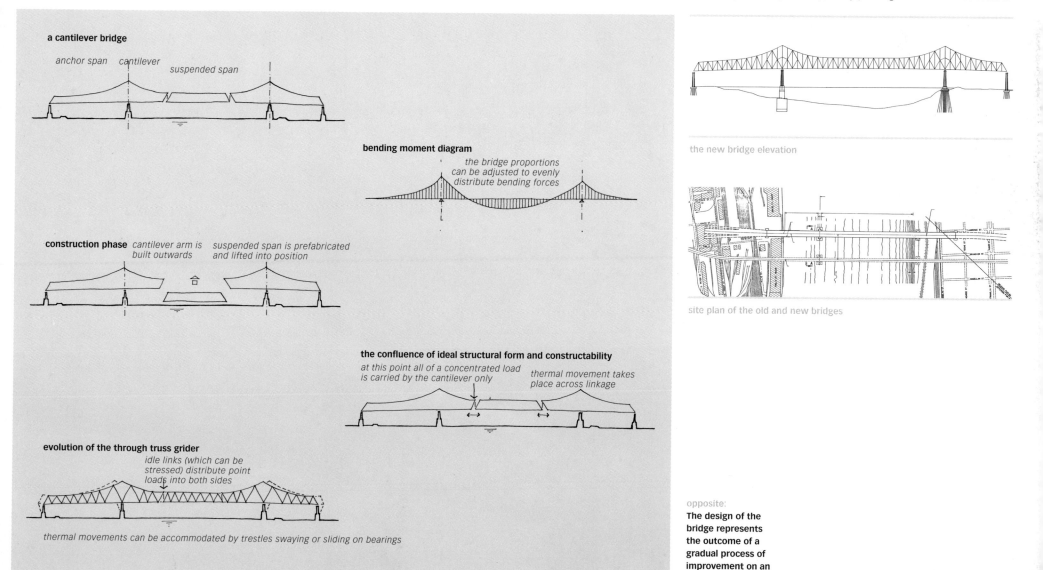

a cantilever bridge

anchor span   cantilever   *suspended span*

bending moment diagram
the bridge proportions can be adjusted to evenly distribute bending forces

the new bridge elevation

construction phase *cantilever arm is built outwards*   *suspended span is prefabricated and lifted into position*

site plan of the old and new bridges

the confluence of ideal structural form and constructability
at this point all of a concentrated load is carried by the cantilever only   *thermal movement takes place across linkage*

evolution of the through truss grider
idle links (which can be stressed) distribute point loads into both sides

thermal movements can be accommodated by trestles swaying or sliding on bearings

opposite:
The design of the bridge represents the outcome of a gradual process of improvement on an established standard.

ice to temporarily alter lengths. It was a short step then to actively stress the final linkages by jacking. The force pattern in the completed structure could be adjusted to the ideal and confirmed in the process.

Thus, by a series of short innovative steps the cantilever bridge had been transformed into a modern through-truss girder, a single integrated form distributing steel stick elements into a highly efficient space structure using the minimum of material.

The widespread adoption of this bridge form was further aided by its clarity and amenability to easy analysis procedures, particularly graphical statics. In this procedure simple vector diagrams will only close if the internal force pattern is balanced by the external applied loads, a self-checking design method contributing directly to safety.

Early on it was realized that it should be possible to work out the ideal proportions for the whole and parts of such structures, for example the relationship between main and side spans and the relative height over the supports to a given span. The logical conclusion of these speculations came in 1904 when the Australian mathematician and engineer Anthony George Maldon Michell (1870–1959) set out the conditions for optimizing all plane frame structures. Twelve years later the American engineer John Alexander Low Waddell (1854–1939) could still claim in his influential textbook *Bridge Engineering* (1916) that through-cantilever bridges were less efficient than simple trusses, an error tempered by his assertion that: 'for some years there has been a slight tendency in American cities to

build cantilever bridges because of their novelty and, possibly, because the city fathers were inveigled into adopting them by smooth-tongued bridge agents,' a rare insight into the breadth of influences on bridge procurement.

The Second Greater New Orleans Bridge could be seen to represent the late phase of this tradition. Sited next to a 1958 bridge of similar form, the new structure is wider and slightly higher than its neighbour. Improvements in the assessment of the buckling capacities of slender elements allows the later design to be rather more spread out in space. The lifting equipment now available makes it possible to place heavier members and their connector plates in higher positions.

The technical advances embodied in the new project are distributed across several fronts. In the design the so-called 'load factor' method was used for the first time on a bridge truss. The method relies on different safety factors applied to different conditions, materials and forms of construction. Dead loads can be more accurately predicted than wind loads and so attract a lower safety margin. Cost savings become apparent as the design is built up. The approach marks an early step towards the introduction of real probabilistic methods, where the assessment of the chances of failure will become the norm and inform future bridge designs.

Tie bars are critical elements of truss structures – their failure is catastrophic. Fatigue cracking is a recurrent concern in steel bridges.

Embrittlement in heat-affected areas around welds and the stress concentrations that occur close to pin holes have been carefully studied by the designers of the Greater New Orleans bridge to find ways to mitigate the problem. The fatigue resistance of weld groups and member shapes has been significantly improved through careful detailing.

Construction methods have been refined to the point where a central section is no longer prefabricated but the two sides are simply extended to a central closure. The partially cantilevered stage of construction becomes critical to the design. At closure the structure goes from one pattern of force conditions to another. In the construction of this bridge, scissor jacks set into the top chords brought the sides together and then allowed the ideal stress pattern to be set before completion.

The firm responsible for the design of the Greater New Orleans bridge was founded by Ralph Modjeski (1861–1940), one of the pioneers of modern American bridge-building. It is worth quoting part of the firm's mission statement to illustrate a sensibility that runs parallel to contemporary preoccupations with innovation: 'At Modjeski and Masters, the future will always be an extension of the past. Our work is influenced by a legacy of evolutionary designs, long-lasting structures, attention to detail, and a belief that a small part of the human soul goes into each project.'

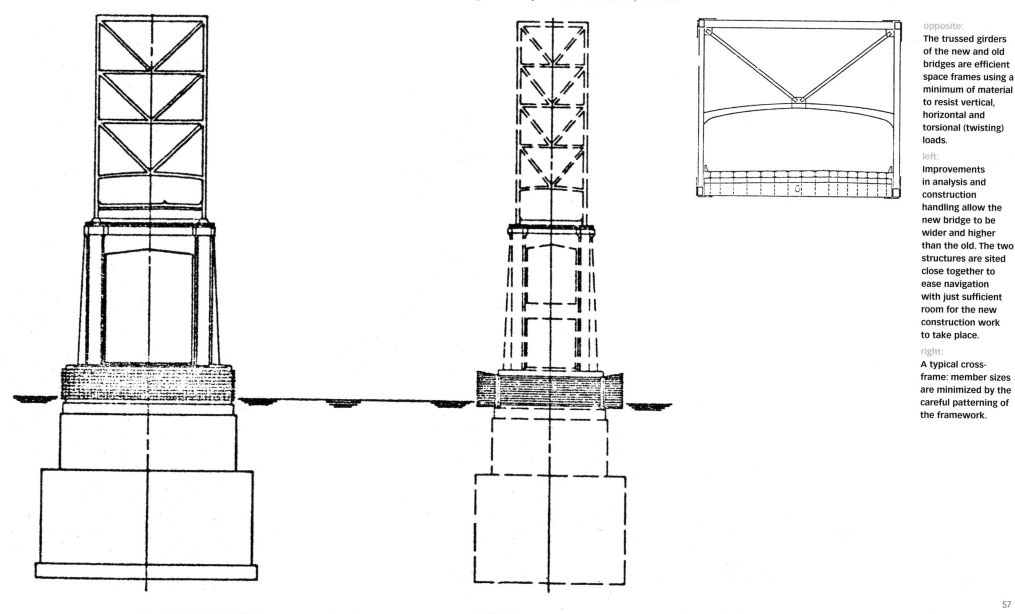

opposite:
The trussed girders of the new and old bridges are efficient space frames using a minimum of material to resist vertical, horizontal and torsional (twisting) loads.

left:
Improvements in analysis and construction handling allow the new bridge to be wider and higher than the old. The two structures are sited close together to ease navigation with just sufficient room for the new construction work to take place.

right:
A typical cross-frame: member sizes are minimized by the careful patterning of the framework.

Campo Volantin Footbridge; Bilbao, Spain; Santiago Calatrava 1990–97
71m / 233 ft

The many pastiches of the Spanish architect-engineer Santiago Calatrava's bridge projects show just how much of an art bridge design has become. Bridge-building could, perhaps, be said to cross the divide between science and art, benefiting from a steady accumulation of knowledge but also requiring the unique expression of the individual. For instance, the insight of the German engineer Fritz Leonhardt led to the introduction of aerodynamically stable bridge cross-sections into modern bridge construction, whereas, through

Calatrava, a succession of sculpturesque bridges has been generated and explored.

Calatrava's work is unique and idiosyncratic. As a result of his designs a new awareness has developed, but not a coherent set of premises for general use, although imitations may reveal the phenomenon of 'not Calatrava', under critical analysis. His Catalan background, with its tradition of building plastic form, and his own experience of working for some of the great Swiss master engineers,

have combined to produce a mixed bag of innovation and formal investigation. Some works are fluent, while others seem overblown; all of them are immensely popular and accessible.

His complex method incorporates the belief that his forms are rigorous embodiments of internal force systems, and yet it is only by setting up a form that a strain field can be generated. He believes patterns of force are not pre-existing, to be freed or brought to light by the engineer. His inspiration might come from sketches of birds in

flight or other shapes of movement, drawing on the similarities between the trajectory of a bird, for instance, or the curving arc of a thrown ball, and the repose of an arch or suspension chain. The very different physical systems – projectile and bridge profile – produce similar patterns. A free body in a gravity field describes a parabolic arc as it turns back to earth, and a static catenary takes the inverse path back up to its supports. Similarly, the reflection of a hanging chain is the ideal arch profile. The arch reifies movement in space.

In addition to the frozen dynamics of these funicular forms there is an appeal to the kinesthetic sense. The schemes are presented in Calatrava's literature with an associated generating sketch, and you can visualize the artist's charcoal, moving freely across the paper, in the sweep and movement of the finished form.

The pedestrian bridge across the River Nervion in Bilbao links a popular riverside promenade and park area with a derelict warehouse complex designated for redevelopment as a residential and shopping

quarter. The single span comprises a deck curving on plan suspended from an inclined arch. The bend follows the natural movement of pedestrians approaching from downstream, and this horizontal bow opposes the tilt of the supporting arch so that they create an equilibrium. Applied loads lock the system tighter together and increase its stability.

The conventional arrangement of an asymmetrical arch bridge places the deck in front of the arch. In this variation the arch straddles the walkway and splayed piers of hangers stabilize the main rib, which

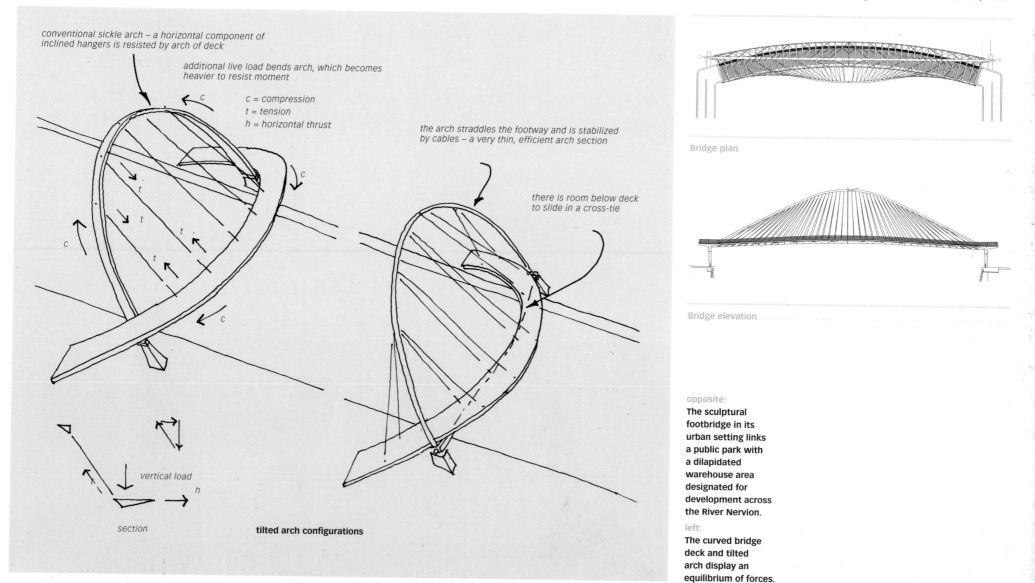

conventional sickle arch – a horizontal component of inclined hangers is resisted by arch of deck

additional live load bends arch, which becomes heavier to resist moment

c = compression
t = tension
h = horizontal thrust

the arch straddles the footway and is stabilized by cables – a very thin, efficient arch section

there is room below deck to slide in a cross-tie

vertical load

section

tilted arch configurations

Bridge plan

Bridge elevation

opposite:
The sculptural footbridge in its urban setting links a public park with a dilapidated warehouse area designated for development across the River Nervion.

left:
The curved bridge deck and tilted arch display an equilibrium of forces.

59

is reduced to a tube of fine proportions. The record of the design's development shows the arch becoming less dominant and the geometry becoming increasingly complex.

Along the inside of the curving deck the array of lightweight hangers describes a graceful envelope of changing angles as each hanger runs between straight and curved margins, a device repeated on several of Calatrava's bridges. Along the outer rim, however, the walking area has to be pulled back from under the raking hangers to provide adequate headroom, leaving a row of unsightly outriggers protruding forward. Furthermore, the cables splay out on elevation and their virtual focus imparts a wheeling movement at odds with the trajectory of the arch. The force system generates a rib with a curved crown and straight sides, with the cables bunching towards the centre section, resulting in an awkward composition. Bridge abutments spread the point loads generated by the bridge ends evenly into the surrounding ground. They also form the threshold between the site itself and the abstract space of balance and resolution occupied by the bridge. The steel central section is set up on shaped abutments of reinforced concrete, while the piers form themselves into approach ramps parallel to each bank. Regrettably, the balance of light steel super-structure on the cantilevered abutment ends is strained, like a sculptural toy, and the bridge touches the ground uncomfortably. Many of Calatrava's bridges are set up on their abutments, rather like a cake on a cake-stand, possibly the outcome of using small-scale model studies.

The footbridge is white, like all Calatrava bridges, and its north–south orientation allows the diffuse Basque light to animate the layering of formal elements in continually changing patterns of shadow. The deck is made from cast-glass panels silicon-sealed into stainless steel rib supports. These translucent units are lit from below to create a theatrical effect that exploits the stillness of the languorous river to generate a dramatic reflection. At night the special lighting makes an elegant pattern in the midst of the lights of the surrounding town.

The sell-out world-tour exhibitions of Calatrava's work have a distinctly hermetic feel, seemingly closed off to other influences. Taken as a group, the models record an exploration of types, experiments in form or mannerist investigations in which changes are made for novelty and to generate consequences for future study. The logic of each bridge is revealed through its precedents, so that the body of work appears to be more important than the individual project. This approach can be contrasted with the work of contemporaries such as

the Czech designer Jiri Strasky, who, hampered by the flat economy of the Eastern Bloc, compensated for the lack of building opportunities by honing a single concept, the stressed ribbon bridge, to complete formal and technical resolution.

opposite:
Glass deck panels and cable envelopes define the walkway space.

left:
The slender arch is stabilized within the weave of cables. Anchorages are concealed to emphasize the slenderness of the structure.

Wettstein Bridge, Basel, Switzerland, Santiago Calatrava, 1988
66 m / 217 ft

This understated bridge by the Spanish designer Santiago Calatrava is also, perhaps, his best. Its design is constrained by the requirements of the site and brief for a new superstructure on existing piers, and is developed as a classic type, an urban roadway across a fast-flowing river, the Upper Rhine. The deep, revetted walls of the water course, built to accommodate spring floods, allow plenty of room for an open-work supporting structure, which is exploited to give an elementary balance to the composition using three equal, arched bays. Two other Calatrava designs were built in the same year that this design was

produced: the clearly stated and well proportioned Mérida Bridge, in Spain, set up near a Roman bridge, and the less well resolved Leimbach Footbridge, in Germany. Both suggest a reaction to his highly wrought and structurally inefficient scheme for crossing the Seine at the Gare d'Austerlitz, in Paris, and to his extravagant project for the Seville Expo, of the year before. Given the profusion of bridge schemes undertaken by this designer, the Wettstein project must be viewed as part of a broader development, and yet it also forms an exception to his body of work.

The succession of white arches and sculptural forms in Calatrava's designs are extraordinary, even moving. They are also unusual in that the completed examples retain their sculptural integrity, and look like very large models. The designs rely for their effect on expressing an internal resolution of forces, and on the curious parallel between dynamic and balanced static forms. The lines of force within a structure follow similar trajectories to moving masses within gravitational fields. By making these invisible lines an intrinsic part of the bridge form, the structure's action is revealed.

In the Wettstein project very little overt reference is made to earlier examples of the designer's work, and the characteristic organic influences are kept low key. Instead there is a strong suggestion of a middle-European bridge-building sensibility: the pylons, classical base proportions and steel detailing are reminiscent of Viennese and perhaps Parisian predecessors.

The structural arrangement is standard. The main supporting arches carry all the bridge's own weight, while their splay reduces the volume of the wide bridge and allows for a shallow deck section in which the sag across the central roadway is balanced by cantilevered promenades on each side. Struts run up to the reinforced concrete deck plate.

Following the tradition of the greatest Swiss arch-builder, Robert Maillart, the arch ring reads independently of the other, secondary elements. Longitudinal stiffness, which is essential in order to spread concentrated loads evenly into the arches' curving lines of thrust and, in turn, to brace the thin compression rings, is provided by a pair of tetrahedral girders beneath the roadway. A subtle secondary curve is given to the bottom booms of these elements to point up the main arch profiles.

This design is not, however, simply a quotation of historical form or a return to the sensibilities of predecessors. The separation of mechanisms embodied in Maillart's schemes was an analytical device to make his structures amenable to the design tools he had available. In Calatrava's work a similar result stems from utterly different processes, in which an initial sketch and model are immediately followed by a computer analysis, almost unlimited in scope as to what

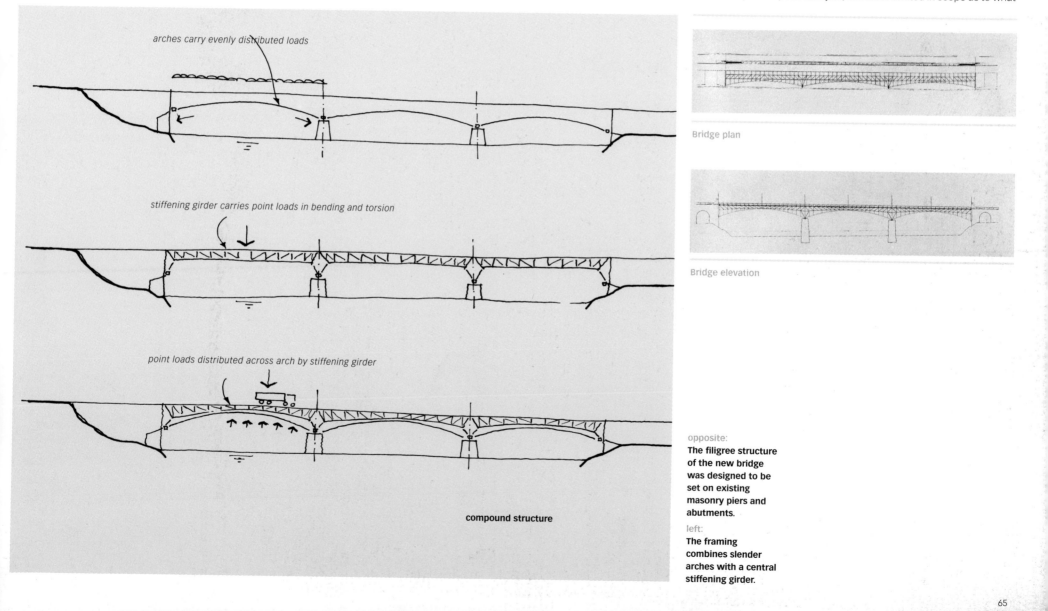

arches carry evenly distributed loads

stiffening girder carries point loads in bending and torsion

point loads distributed across arch by stiffening girder

compound structure

Bridge plan

Bridge elevation

can be studied and resolved. The Wettstein Bridge may look like a *fin de siècle* structure, because it is a similar diagram of forces, but it cannot be analysed so readily. Like so many of Calatrava's bridge projects, it forms part of a line of formal experimentation: the limitations imposed on earlier generations of engineers have been taken up and their implications meticulously worked through.

The trussed upper works of the proposed bridge sit lightly on the old footings of the earlier structure, its filigree of tubular metal contrasting with the heavily rusticated masonry starlings. The cross-section is sophisticated and a process of refinement and simplification can be traced through a sequence of design drawings.

The thickness of the deck plane is reduced to a minimum and the balustrades are open stressed-wire railings, angled to prevent them competing with the deck edge and to preclude climbing. A concentration of piped services is set centrally below the roadway, out of view, and the road is edged with trief curbs: a profiled rim that directs swerving trucks back into the roadway and off the adjacent cycleway. The upper booms of the stiffening girders are set within the shaped soffit and picked out by lights concealed in piercings through the deck, while concave outer soffits catch uplights and reflections off the water. The precast-leading edge of the deck is given a second concave profile, building up towards a classical cornice.

The junction of new and old is, however, handled in a rather self-conscious manner. The deck oversails the piers and the splayed supporting arches are carried on sculpted extensions of the upper pier corners. These form awkward assemblages of intersecting planes; concave surfaces and arrised edges highlight the difficulties of visually

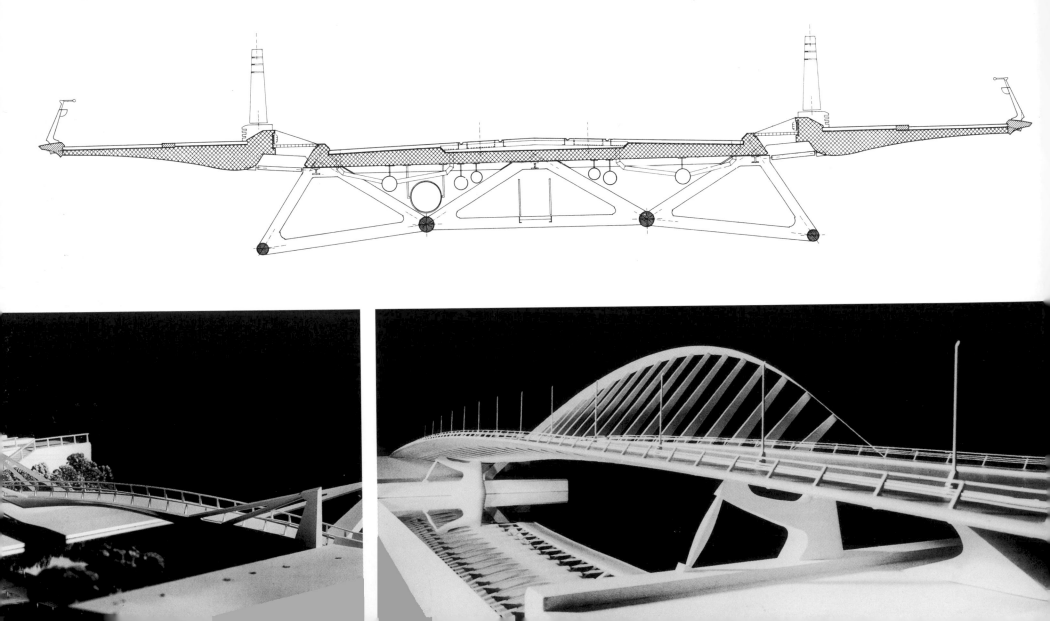

resolving these joints. A sequence of drawings reveals the designer working through a variety of options for these important junctions.

In general the bridge's details reflect less about its construction, the way in which the parts are fitted together, and more about how forces move across the joints, and how each joint's influence radiates back out into the strain field (the structure as a whole). The Wettstein Bridge was first designed to include welded joints, which are visually clear and allow forces to flow evenly across the filleted nodes. The site welding of so many connections is, however, extremely difficult and,

in addition, locked-in stresses bend the finer welded elements off line, severely compromising their axial strength. Pinned joints were therefore substituted in the scheme because each pin guides forces into alignment across its access. The multiple intersections result in the plates having to spread out to allow room for the pins to be set, which is an unfortunate compromise.

Indications of how the structure is put together are otherwise generally suppressed. The deck is moulded and envelopes the supporting truss, a prefabricated, pinned-up assemblage. This overlaying

of two construction forms is ambiguous; it aims to engage the viewer's attention and helps to create an overall effect of immense sophistication. Many influences have been allowed to make their mark in this structure, and the inclusion and resolution of conflicting demands demonstrate both this designer's fluency and the potential of bridge design to produce something more than a structurally expressive artifact.

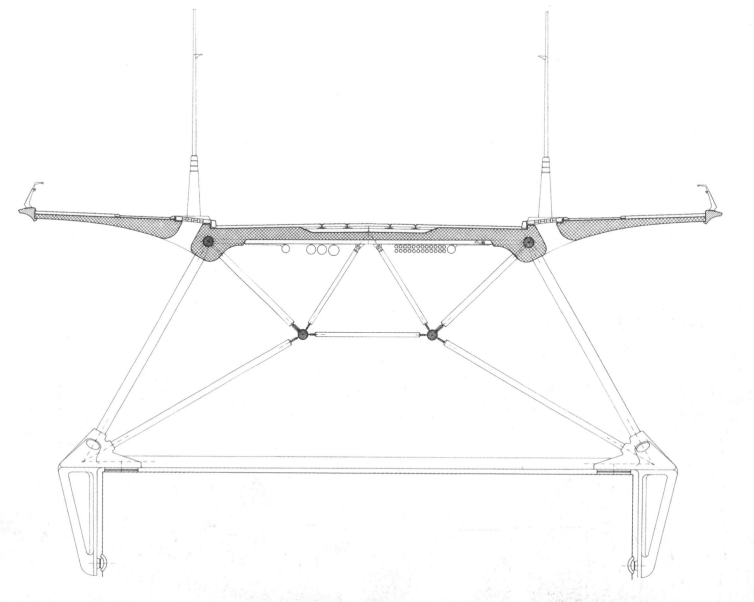

opposite, above:
The bridge cross-section was initially developed with cast nodes and fully welded joints.

opposite, left:
The Leimbach Footbridge scheme, designed in the same year as the Wettstein project.

opposite, right:
The Gare d'Austerlitz bridge, Paris. Another project proposal from 1988.

left:
The final design section uses simple connector plates. The pier bearings are complicated steel castings.

Traversina Footbridge; Viamala, Switzerland; Conzett, Bronzini, Gartmann AG 1999
48 m / 157 ft

The Traversina Footbridge, which was carried away in a landslide in 1999, represented one element in an 'eco-museum'. The product of an organized encounter with a particular environment, it formed part of the landscape's historical interaction with man.

The Swiss Alps form a type of palimpsest, in which a network of old Roman paths is layered with subsequent history, with castles, churches, monasteries and trade routes. Any change to this landscape calls for the least possible intrusion, to prevent the loss of its unique atmosphere. A very light structure was therefore proposed along the ancient Via Traversina, to divert a mule track across the chasm through which the Traversina Torrent flows. The footbridge made a quiet contribution to the surroundings, its details inviting intimacy and recalling the bridge's proximity to nature.

The inaccessibility of the site led to the 48m (157ft) span being lifted in by helicopter. Timber was the chosen material, with the framing held together by cable ties and metal joints. The appropriateness of the renewable resource, and its immediate resonance with the coniferous precipices nearby, were underpinned by another sensibility. The structure had been analysed, developed and detailed in the modern conceptual space but might have been made at any time during the ancient road's existence. The elements could have been bound together and the metal parts blacksmithed. It was a Roman demountable bridge reinterpreted through modern experience, yet it was clearly not a pastiche.

To save weight, a mixture of larch and Douglas fir was employed, species whose strength-to-weight ratios eclipse those of steel or carbon fibre. The drawn-wire ties are an exceptionally strong form of

steel, for which the wire is pulled through a die and then 'work hardened'. The attendant brittleness is mitigated by the twisted-strand structure of the rope. Each fibre is delicate but the whole cable is highly robust, resilient and supple, forming the finest possible filigree.

With a lifting limit of 4.3 tonnes (4.23 tons) the structure was assembled in two pieces, having been conceptualized in three parts. The main load-bearing element was a simple three-chord truss: the upper member was a laminated strut of larch, the lower chords were each continuous, twisted-strand cables. The bridge's parabolic profile

was the ideal geometry and therefore kept stresses even throughout the length of the lower girdles. This is a favourite device among Swiss German engineers and grounded in the beautiful lenticular forms of Friedrich von Pauli. A continuous beam of laminated timber, placed horizontally to resist wind loads, was set onto the truss. To reduce twisting in the wind, the beam level was carefully set to be near the centre of pressure along the bridge sides. The walkway formed a separate timber deck above. Solid side balustrades, reassuring for pedestrians standing at the airily exposed centre of the bridge, were

locked to the horizontal plate and supporting truss by H-frames at regular intervals. The resulting channel section resisted torsional forces, keeping the deck level under uneven loadings.

Each part, including the truss, horizontal beam and deck channel, was strong enough to carry its own weight and the temporary pattern of forces set up by the air lift. Locked together the whole assembly could carry the imposed loads of walkers and wind. The awkward junction of the H-frames and the trussed base was a compromise between serviceability and construction requirements. Successive

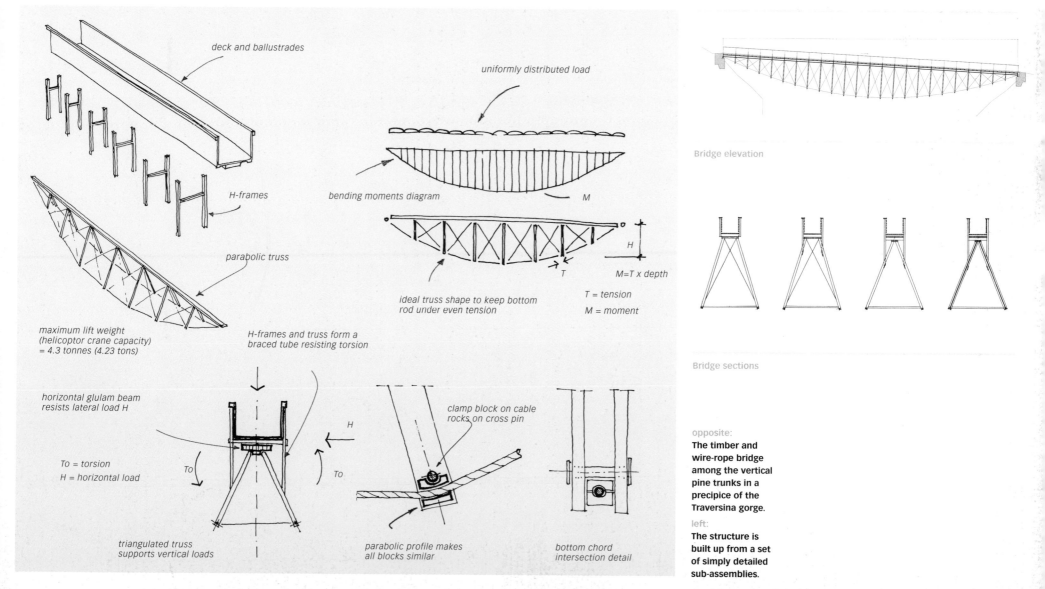

deck and ballustrades

H-frames

parabolic truss

maximum lift weight (helicoptor crane capacity) = 4.3 tonnes (4.23 tons)

H-frames and truss form a braced tube resisting torsion

horizontal glulam beam resists lateral load H

To = torsion
H = horizontal load

triangulated truss supports vertical loads

uniformly distributed load

bending moments diagram

M

ideal truss shape to keep bottom rod under even tension

H

T

$M = T \times depth$

T = tension
M = moment

clamp block on cable rocks on cross pin

parabolic profile makes all blocks similar

bottom chord intersection detail

Bridge elevation

Bridge sections

opposite:
The timber and wire-rope bridge among the vertical pine trunks in a precipice of the Traversina gorge.

left:
The structure is built up from a set of simply detailed sub-assemblies.

sections in the design development show the supporting truss becoming independent of the deck and sidings above, emphasizing the interface between the two parts.

The detailing demonstrated a profound understanding of timber structure. All the parts were built up from economic cuts. In a trussed frame the concentrated loads at each end have to be transferred from member to member. These forces must be spread into the soft and compressible grain of the fast-growing coniferous timbers. Tooth-plate connectors improve the strength of bolts by dispersing stresses outwards. Timbers can be doubled around steel 'flitch' plates so that

loads can be concentrated into the metal and back out again. The surrounding timber protects and supports the thin steel sandwich plate. The most difficult detail on this bridge, the bottom tie to the upright connection, was beautifully resolved by a simple rectangular clamp block on the stainless steel cable (a particularly safe option because of the very high coefficient of friction of this material). The block was carried on and rocked around a connector plate set into a strut built up from four sections.

In Scandinavia and North America the longevity of wooden building has allowed sophisticated uses to be developed. South of Scandinavia

the European climate has higher relative humidity (more than 20 per cent) and protection therefore becomes a preoccupation. Moisture in the air supports wood-boring insects and rotting fungi. Cut timber exposes end grains, where the pores of the wood formed to carry sap in the living organism; these are particularly susceptible to moisture. The rotting-tooth outline of old marine piles and wrecked ships' ribs is typical of the havoc that can be wrought. Principal means of protection involve casing the timber elements. On the Traversina Footbridge the upper strut ends were carefully arranged to intersect with covering elements. The heavy horizontal beam was covered with

bituminous felt and copper edge drips to form a roof over the main load-bearing elements. Secondary means of protection involve sealing the wood or impregnating it with poison. For the latter, copper chrome arsenic is driven into the timber grain under pressure, leaving slightly green pieces of wood to weather more slowly to the familiar silver-grey. The obvious ecological downside of this practice was mitigated at Traversina by only treating those elements that cannot be replaced at some future date. The deck and balustrades, with which pedestrians can come into contact, were all left untreated. The structure exploited the drop into the gorge below to achieve the maximum depth and, therefore, the maximum stiffness. The assemblage of so many parts gave the system a very high degree of internal damping; thousands of joints in the structure rubbed against one another, fretting and shedding energy, as each impulse passed. Timber itself provides a good deal of hysteresis damping. Every time it is flexed or strained some energy is lost as heat within and between its fibres. The bridge was, therefore, inert to footfall vibration. The confined surroundings meant that the structure was buffeted by choppy gusts of wind but a steady build-up of dynamic response in an airflow was precluded by the aerodynamic damping inherent in the box-kite form. The bearings were simple steel rollers on half jointed timber blocks, a sign from one engineer to another that the overall intention has been the expression of studied simplicity, the Holy Grail of the engineer.

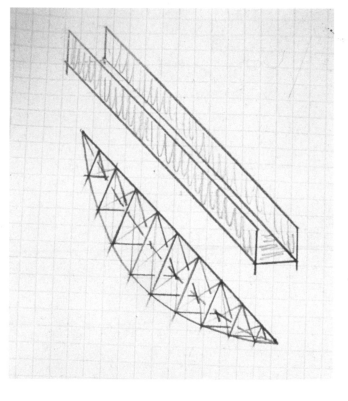

opposite:
The structure was eventually destroyed by a landslide in 1999.

left:
The man-made object brings a taming influence into a wild landscape designated an 'eco-museum'.

above:
The designer's sketch for the prefabricated components that were lifted in by helicopter.

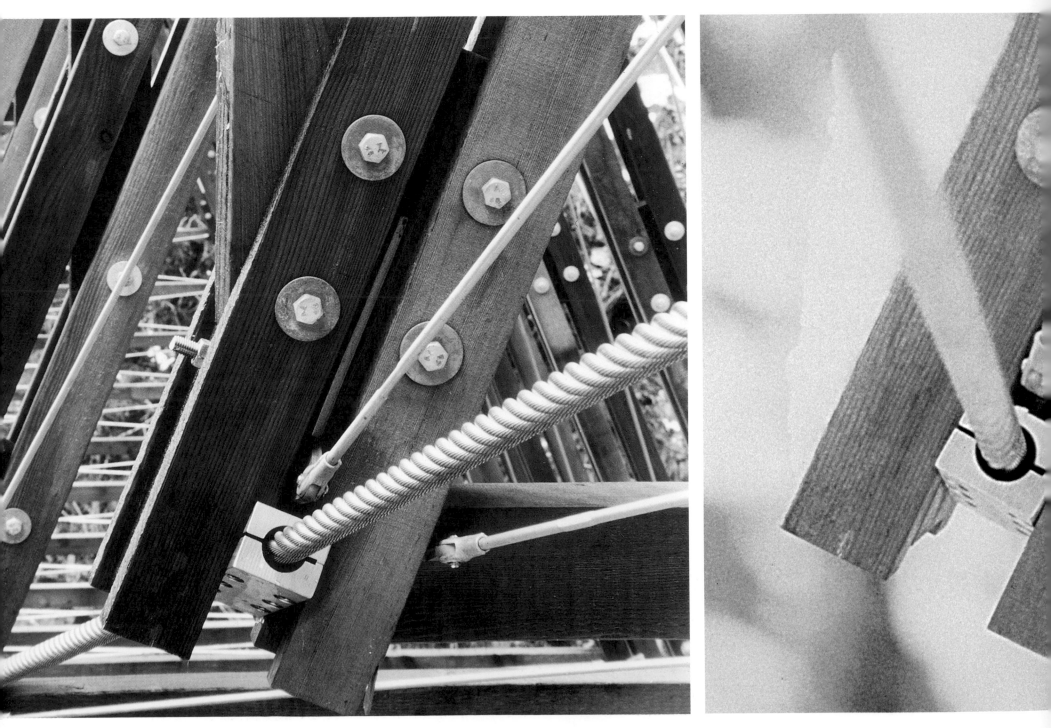

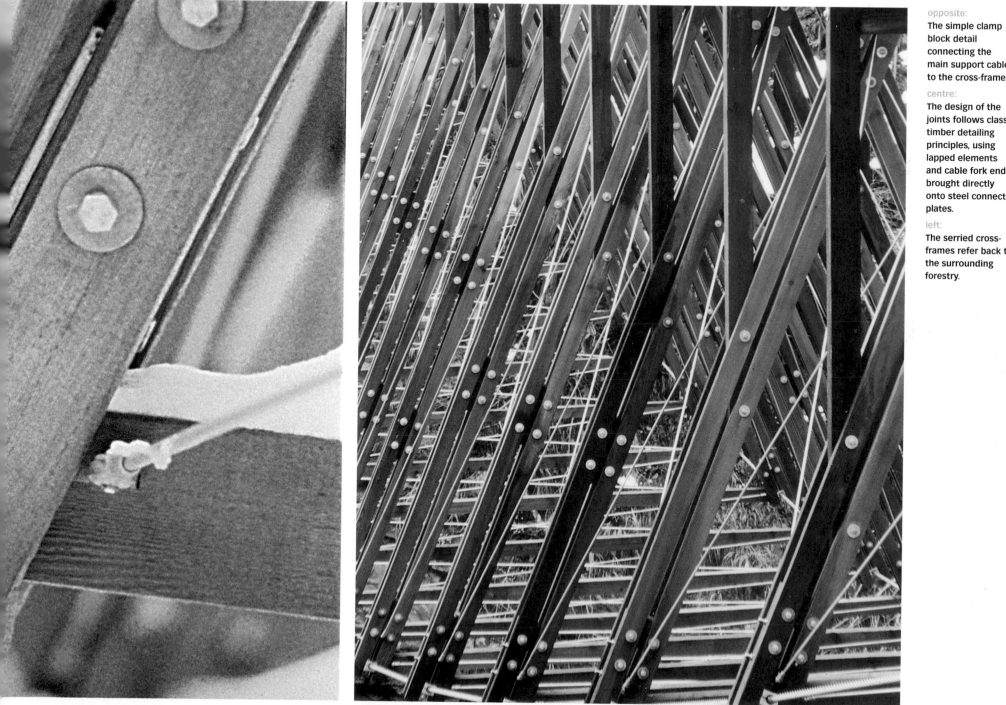

opposite:
The simple clamp block detail connecting the main support cables to the cross-frame.

centre:
The design of the joints follows classic timber detailing principles, using lapped elements and cable fork ends brought directly onto steel connector plates.

left:
The serried cross-frames refer back to the surrounding forestry.

Pùnt da Suransuns; Viamala, Switzerland; Conzett, Bronzini, Gartmann AG 1997
40m / 131 ft

When first encountered, the Pùnt da Suransuns makes a profound impact, suggesting as it does the most complete solution to a bridging problem. There is nothing here but a walkway, handrails and space.

Its apparent simplicity has, however, been striven for. A seemingly straightforward structural mechanism has been employed and subtly manipulated with the utmost sophistication. The details clearly reflect their purpose; the tectonics are elemental, almost Shaker in their conviction and in their elevation above the practical.

The scheme, which is the result of a competition open to local engineering firms, is rooted in the site. The level change from one end to the other is just enough to make the catenary work. The form is unadulterated, a pure line among the jam of boulders in the valley bed.

The structure comprises a deck of stone slabs supported on straps of high-tensile Duplex steel hung between anchor blocks. The two parts are stressed together to act compositely. The irreducible clarity of the evenly loaded catenary, described by the most elementary and

elegant of differential equations, ramifies in complexity when stiffening is added to account for uneven forces. If a point load is applied to a rope it changes shape. Deflections do not vary in a linear manner but are magnified by the distortions. The essential strategy in a ribbon bridge is to take the concentrated load, in this case an individual or a bunched group of walkers, and redistribute it as widely as possibly across the suspension chain. This avoids the development of a sharp 'corner', where high bending stresses and deflections will

occur. Here the stone planks are compressed together to act as a local bending element rather like a strongman holding together a row of bricks in the air.

The design centres on the studied reduction and simplification of the constituent parts. This is 'Ockham's razor' at work: the medieval idea that the hidden reality of natural forms has a transcendental simplicity. Reaching the truth, or doing something appropriately, involves paring away until only the essential remains.

Stone planks used for the deck form a tacit continuation of the ancient roadway served by the bridge. The blocks are compressed together from each end. Load-transfers between the stones are made not with mortar but with ductile aluminium strips, which yield around hard spots, are corrosion proof and provide reinforcement to the flawed material. The walking surface of the granite is flamed to reduce the likelihood of slipping. The main support straps of the bridge are flat bars of Duplex, a high-strength steel of exceptional durability

developed for the chemical and food-processing industries. Each side has a double strap that fits neatly around the anchor block, reinforcing this critical element for greater safety.

Visual and physical intrusion is kept to a minimum in the balustrade, which consists of a set of uprights with a top rail and is unclimbable. Each balustrade post passes through the centre of a lintel, between the supporting straps, beneath which it is threaded to take a back plate and lock nut. The posts are both supported by the

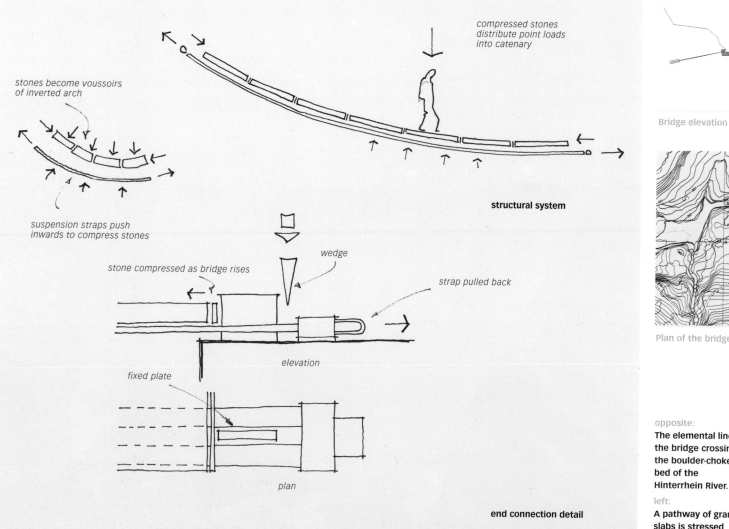

compressed stones distribute point loads into catenary

stones become voussoirs of inverted arch

suspension straps push inwards to compress stones

stone compressed as bridge rises

wedge

strap pulled back

structural system

elevation

fixed plate

plan

end connection detail

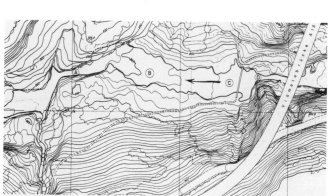

Bridge elevation

Plan of the bridge and surrounding gorge

straps and stones and pull them together. As each strap approaches the abutments it runs over a build-up of layers, forming a leaf spring. This bearing responds with progressive firmness, reducing the secondary effects that the governing equations predict will develop at the less than ideal boundary conditions.

At their terminations the main straps run along either side of a simple fin plate cast deep into the anchor plinth. End blocks have hasps welded onto them. During construction the straps were hung off these staples onto turn-buckles tied back on the anchor blocks. The stones were then set in place bearing tight up against the inside of the fin plates. Tightening the turn-buckles lifted the sagging flats, compressing the train of slabs against one another into a locked whole. Finally steel folding wedges were driven in between the strap end blocks and back face of the vertical fin plate, an index of what has gone on, and the stressing assembly removed. The ground anchors to the abutments have exposed heads set back beyond the bridge assembly. All parts of the structure drain freely, are visible for inspection and readily replaceable.

The relentless resolution of the design stems from a more profound background than the studied care of minimalism. Chur, in the Graubünden canton of eastern Switzerland, has a particular sensibility to building: an affinity for materials, a respect for peasant 'ways of doing' combined with a cultured modernity and a distinctive mix of Latin and Germanic influences. The insularity of the mountains

leads to an introspective and hermetic viewpoint that is free from any sense of universality. The designer, Jürg Conzett, trained as a civil engineer before completing an apprenticeship with the celebrated architect Peter Zumthor, whose craft background, close attention to materiality and soft modernism has proved accessible and popular. A critical mass of such designers seems to have been reached in the area. Their products are immediate and timeless. The continuity they represent is reflected in the immediacy of the words of Adolf Loos, an early modernist architect from across the nearby Austrian border. In his essay 'Rules for those building in the mountains', of 1913, he states: 'Build as well as you can. No better. Do not outstretch yourself. And no worse. Do not deliberately express yourself on a more base level than the one with which you were brought up and educated. This also applies when you go into the mountains. Speak with the locals in your own language. The Viennese lawyer that speaks to the locals in a country bumpkin's accent is beneath contempt. Pay attention to the forms in which the locals build. For they are the fruits of wisdom gleaned from the past. But look for the origin of the form. If technological advances made it possible to improve on the form, then always use this improvement....

'Be true! Nature only tolerates truth. It copes well with iron truss bridges, but rejects Gothic arched bridges with turrets and defensive slits.'

opposite, left:
The bridge serves as the continuation of an ancient Roman paved road.

opposite, right:
The balustrade uprights secure the deck stones to the main supporting straps.

top left:
The abutments integrate ground anchors and the bridge stressing system within one large cast block.

bottom left:
The strap ends and wedges at the supports are left exposed, and the method of construction is visible to the passer-by.

The simplicity of the bridge's form means that intrusion into the landscape is minimal.

The level change across the site is sufficient to provide the required catenary without excessive sag.

The Great Belt Link (East Bridge); Great Belt Strait, Denmark; Dissing and Weitling/COWIconsult 1998
1624 m / 5328 ft

The largest man-made structures, whether city buildings or civil engineering works, change the landscape. At a specific scale an artifact is no longer contained by but instead reconfigures its surroundings. Despite the huge resources required for such structures, and their longevity, strategies to control their impact have evolved slowly.

One particularly positive example is found in the East Bridge road and rail crossing, the final link in a chain of three crossings over the 18 kilometre- (11 mile-) wide Great Belt Strait. The integration of all aspects of the project from the start reflects Denmark's sophisticated environmental awareness. From the overall layout to the details of structural junctions the project demonstrates a consistency of approach. The project designers incorporated the functions of engineering, architecture and landscape architecture, environmental planning, product design and graphics into one tightly managed team.

The site of the crossing was determined by Sprogø Island, which divides the strait into east and west channels of similar size, both of which could be spanned by tunnels or bridges on a scale already completed elsewhere.

The landscaping of large civil-engineering structures has often involved adjusting the form to reduce its impact on the surroundings, to camouflage the installation or to make the object appear more 'natural'. For the Great Belt engineering complex the architects adopted a subtler strategy. The existing landscapes were analysed for specific characteristics and qualities and then the design developed as a counterpoint, underscoring and interpreting these features. The two sides of the strait are quite different: on the east side, Zealand, the rail approach is set along the base of a fjord and the alignments

respond to the surrounding hills, dunes and outcrops. To the west, the approach alignment and embankment follow local deposition features in the form of long sand ridges and are planted with long, laminar belts of pines.

Adjacent earth structures, embankments and cuttings are not styled as naturalistic landforms, such as those on a golf course, but are set up as clear geometric forms. These huge, simple shapes reflect the scale of the man-made forces involved in moving the earth; they also evoke a sense of the tectonic forces that have shaped the surrounding glaciated landscape, leaving curious, neatly spaced, artificial-looking deposits in the form of drumlins and eskers.

The road and rail links follow a common alignment across the west channel, rising and curving on a low-piered bridge to the island before separating into the large, high east bridge and rail tunnel. The road lines were analysed spatially to provide a sequence of views opening onto the big bridge and to avoid ugly foreshortenings of perspective. The continually changing curvature encourages concentration in drivers and enhances road safety.

The environmental assessment of the area covered the flora and fauna to be maintained but also extended to the tidal movement of the Baltic Sea. It was necessary to find a solution that would maintain the normal flow of water in and out of the strait. The artificial peninsulas required for a road tunnel under the east channel, the shipping channel, would have proved too intrusive and the option of a large bridge was considered instead. The rail tunnel swoops under the bay, curving through a subterranean belt of moraine gravel, allowing for a high, wide and light road bridge above. Extensive analysis of

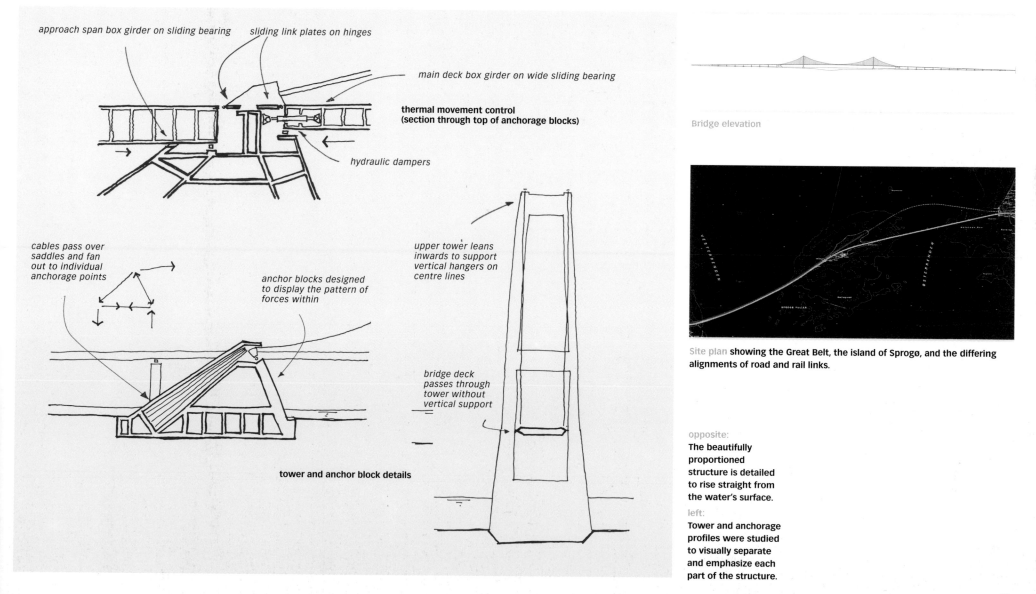

approach span box girder on sliding bearing

sliding link plates on hinges

main deck box girder on wide sliding bearing

**thermal movement control
(section through top of anchorage blocks)**

hydraulic dampers

cables pass over saddles and fan out to individual anchorage points

anchor blocks designed to display the pattern of forces within

upper tower leans inwards to support vertical hangers on centre lines

tower and anchor block details

bridge deck passes through tower without vertical support

Bridge elevation

Site plan **showing the Great Belt, the island of Sprogø, and the differing alignments of road and rail links.**

opposite:
The beautifully proportioned structure is detailed to rise straight from the water's surface.

left:
Tower and anchorage profiles were studied to visually separate and emphasize each part of the structure.

current and future requirements for navigation indicated the need for a bridge with a main span of at least 1600m (5249ft) and therefore a classical suspension bridge scheme, traditionally viewed as the most economic option for spans of over 1500m (4921ft), was adopted.

The consistent geological conditions beneath the water placed the east bridge out in the centre of the channel's expanse. The anchor blocks rise from the sea bed so that the bridge's form is not influenced by any external accidents of location but settles into a repose brought about by its idealized form. The proportions of the main span to the back spans, of the catenary curvature and of height to width are all determined by a structural efficiency that is almost incidentally visually harmonious.

Several architectural devices are harnessed to achieve a palpable monumentality and grace. The deck is an aerodynamically shaped steel box girder formed as a continuous strip between the abutments. It is then visually linked across the expansion joints to the approach spans by maintaining the profile but expanding to a deeper box girder between the piers. The continuity and independence of the deck as it passes the towers is emphasized by the unusual step of omitting the cross pieces, which traditionally sit below the road deck providing support as it passes between the tower legs. Here the suspended element runs straight through, while the unobtrusive lateral supports pick up only wind loads. The visual separation of the deck and the support is maintained on the approach spans, where large gaps are left between each pier and soffit into which bearings are placed and where they can be easily monitored and replaced.

The uprights and crosspieces of the towers were developed as monoliths, rather than as assemblages. Their form resonates with the Corbusian adage: 'architecture is the magnificent play of geometric forms brought into the light'. The load path down through the tower is

clearly expressed. The towers slant outwards at their peaks, picking up the main cables on their centre lines and then passing around the roadway. The caissons on which the towers rest are set 3.5m (12ft) below the water to give the impression that the towers plunge seamlessly beneath the surface. Ship strikes are avoided with carefully graded, submerged sandbanks rather than with extensive and obtrusive starlings or cutwaters. Gravel covers protect the banks from scouring and limit hull damage if groundings do occur, to prevent oil spillages. Suspension-bridge anchor blocks are traditionally monolithic, their sheer weight stopping them from slipping inwards. In

the east strait the forms have been hollowed out and reduced to massive tripods that emphasize the forces contained.

Detailing on the bridge is direct and unfussy. The main hangers meet the deck with sock dampers of studied simplicity. Prestressed wire barriers divide the carriageways so that the view is not interrupted and secondary aerodynamic effects are avoided. Thermal movements are controlled at each abutment by hydraulic buffers in carefully laid out access spaces. The internal cells of the steel box girder are protected from corrosion by dehumidifiers, a Rolls Royce solution to the perennial problem of condensation in hollow metal

components, more usually dealt with by sealing or wax injection. The huge rocker boxes on which the main cables sit are air-conditioned.

The bridge lighting is low-key and subtle. Concealed floodlights wash the inside surface of the towers, hangers and anchorage blocks. Simple points of light along the catenary complete the accentuation of the form without displaying a huge, brightly lit object.

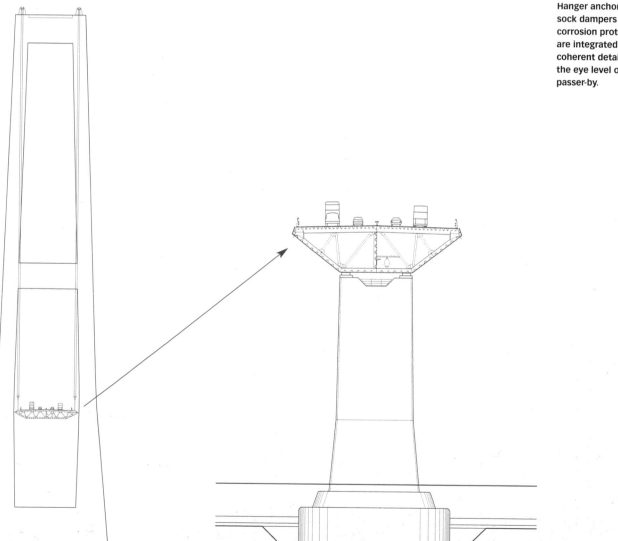

opposite, left:
Hanger anchorages, sock dampers and corrosion protection are integrated into a coherent detail at the eye level of the passer-by.

opposite, right:
Tower lighting emphasizes the inner surfaces of the vertical elements while avoiding excessive light pollution.

left:
The box girder profiles of the main span and approach viaducts are visually related.

opposite:
The subdued lighting scheme picks out the main elements and profile of the bridge.

left:
At 254 metres (833 feet) above water level the tower tops emerge above a sea fog during construction in 1995.

above:
Wooden study models of the main towers, anchorages and approach piers.

Millennium Bridge; London, UK; Foster and Partners – Anthony Caro / Ove Arup and Partners 2000

162 m / 532 ft

The design of the pavilion for the Great Exhibition of 1851, the Crystal Palace, was worked out over lunch on a train journey between Leeds and London. An ink sketch on a linen napkin, preserved in the Victoria and Albert Museum in London, records the design process. The apparent effortlessness of Joseph Paxton's inspiration is, understandably, part of its attraction and it finds a parallel, almost 150 years later, in the pristine, green wine-bar napkin used by Foster and Partners to encapsulate the design intent for London's Millennium footbridge at Bankside.

This crossing of the River Thames between St Paul's and Bankside had been under consideration for some time, particularly once the development of an old power station south of the river into a modern art gallery made the completion of a tourist route across the river imperative. A speculative scheme by the engineer Robert Benaim and the architect Peter Clash had shown how the sensitive prospect of the City of London, on the north bank, could be maintained and enhanced with a low-profile suspension bridge. Their 'stressed ribbon' solution, integrating cables and deck, had difficulty achieving the shallow ramps

necessary for disabled access but it paved the way for a radical solution.

Submissions for the subsequent design competition dealt in different ways with the need to span an extremely shallow depth. One proposed a through-truss with glass components, another a monocoque wing-like structure. The winning scheme, by Foster and Partners, utilized the immense strength of locking-strand cable to make a suspension bridge unprecedented in its transparency and attenuation.

The sketches show renderings of various structural solutions, culminating in three pen strokes labelled 'plan' and 'section'. The

architect's design instruction was, effectively, 'no structure' and the epithet 'blade of light' was used to give dramatic reinforcement. In keeping with this, every means was exploited to make the bridge appear merely as a line in space.

This stretch of the Thames is shallow and so the piers could be placed well out in the stream, allowing the bridge to divide into similar spans. The catenaries were inclined inwards at an unusually steep angle and connected through substantial cross-frames to give the thin deck lateral support. The walkway is set in a graceful profile from side to side with the cables rising and dipping around this centre line. The cables were locked off over the towers to increase overall rigidity. Unfortunately, because the deck section is pared down to the minimum, the main cable arrays must be inflated into heavy bunches to cope with the highly distorted form, which results in the two elements approaching visual parity and conflicting with each other. The description 'blade of light' is appropriate at night, however, when careful lighting emphasizes only the deck.

To speed construction the deck was made in lengths that could be transported by road, and the many elements were then linked but not connected into a continuous whole. Notably, little advantage was taken of the opportunity to introduce additional damping at these points, added to which the open balustrade, an important part of the overall, attenuated appearance, contributes negligible aerodynamic damping. The standard suites of wind-tunnel testing on aero-elastic models had, however, shown that the system was aerodynamically stable.

The Millennium Bridge was inundated with crowds on its opening day but within 48 hours it had closed, after it was discovered that the

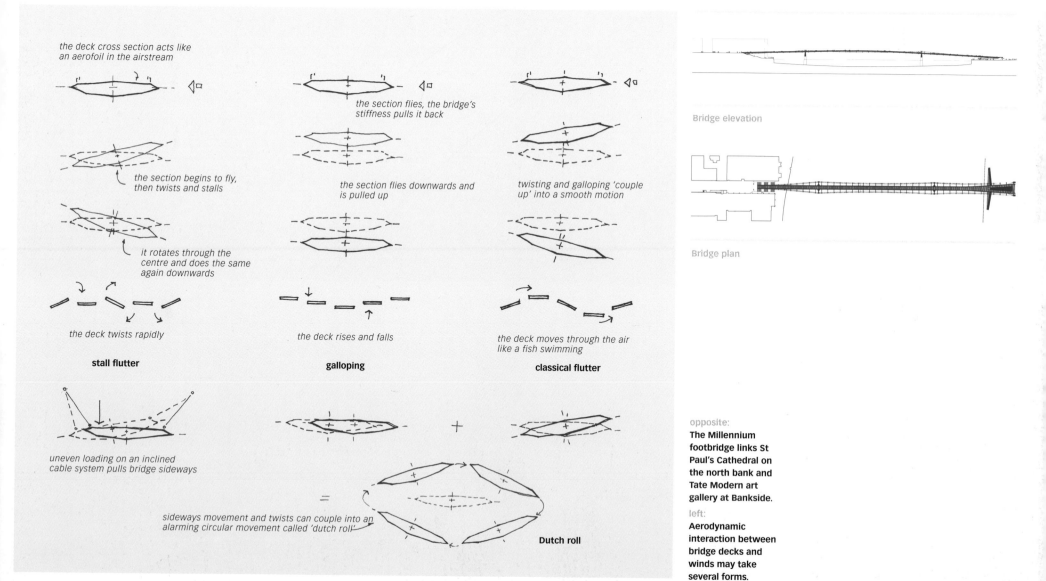

the deck cross section acts like an aerofoil in the airstream

the section begins to fly, then twists and stalls

it rotates through the centre and does the same again downwards

the deck twists rapidly

stall flutter

the section flies, the bridge's stiffness pulls it back

the section flies downwards and is pulled up

the deck rises and falls

galloping

twisting and galloping 'couple up' into a smooth motion

the deck moves through the air like a fish swimming

classical flutter

uneven loading on an inclined cable system pulls bridge sideways

=

sideways movement and twists can couple into an alarming circular movement called 'dutch roll'

Dutch roll

Bridge elevation

Bridge plan

opposite:
The Millennium footbridge links St Paul's Cathedral on the north bank and Tate Modern art gallery at Bankside.

left:
Aerodynamic interaction between bridge decks and winds may take several forms.

structure oscillated under footfalls, and that alarming levels of lateral movement built up when large groups of people fell into step. The south-west side span proved particularly prone to movement.

The now commonplace order for soldiers to break step on bridges – for example it is cast into plaques on the piers of London's Albert Bridge – originates in an incident at Angers in 1850 when a 500-man unit of the French army kept in step while crossing a suspension bridge over the River Saône. A resonance built up until one of the rusty backstay cables came out by the roots, leading to the deaths of 250 men. Although the armies of the world learnt a valuable lesson, one

vital nuance seems to have been overlooked: to achieve this effect the input must be in phase with the oscillation. As with a child's swing, movement only builds up if the push is at the right moment. If there is any slight disparity then the action reduces rather than increases motion. Although it was not realized at the time it seems unlikely that the troop of French soldiers began by marching right on the natural frequency of the bridge or that the men were so perfectly drilled that they kept metronomic beat as the bridge began to sway. It has since been discovered that people walking en masse will lock into the beat provided by the structure itself rather than shy away from it.

Some of the design features of the Millennium Bridge give it unusual vibration characteristics. The inward slope of the main cables towards the deck makes the vertical, the horizontal and the torsional stiffness of the structure comparable. When gravitational loads influence the vertical frequencies, the horizontal and torsional modes inevitably remain very similar. Vertical loads that are out of balance, such as synchronized footfall, cause horizontal as well as vertical movements, resulting in the so-called 'Dutch roll': in this effect sideways and twisting motions couple together, causing the deck to rotate but stay level.

On a continuous suspension bridge the widely differing main span and back spans usually restrain each other; movement in one part cannot transfer to another without mobilizing large damping forces. The parity of spans on the Millennium Bridge precluded this calming effect, however.

Apart from major interventions to improve stiffness, the remedy for footbridge vibration is to damp the structure so that more energy is removed from the system on each cycle than can be put in by people acting in unison. However in this instance a complication arises in that the imposed load, the pedestrians, represent a significant part of the total weight of the system. Frequencies vary with mass, and adding shock absorbers tuned to one frequency will not help when conditions change and frequencies move off elsewhere.

It would have been possible to arrange active damping on the bridge that adjusted, like the suspension of a Williams racing car, to varying conditions but this attractive 'engineer's solution' was denied by time constraints. Instead, stiffening was added and a combination of shock absorbers, which cut down perceptible vibration and prevented people from 'locking on' to the onset of sway, as well as heavy tuned dampers to counter resonance in congested conditions.

A great deal has been learnt from this incident about the protocols of dealing with vibration serviceability on lightweight bridges. It highlights, in particular, the fact that computer models can be blunt tools, and that proper testing and updating procedures for models will necessarily become the norm.

opposite, top:
An earlier 'blade of light' proposal for a stretched ribbon bridge, by architect Peter Clash and structural engineer Robert Benaim.

opposite, bottom:
The long, low curving deck profile aligns with St Paul's Cathedral.

left:
The monumental piers are made up of elliptical sections, reinforced concrete bases cast against steel-faced formwork, and pressed and welded steel plate upper arms.

above:
The walkway section is reduced to the minimum number of functional parts. The deck is extruded aluminium planking.

West India Quay Footbridge; London, UK; Future Systems / Anthony Hunt Associates 1996
15m / 49 ft

The power of bridges – their financial requirements and their ability to transform geography and economics – make them a focus of political activity. At the end of the eighteenth century, for example, it was the very backwardness of the Hapsburg Empire in technical matters that led them to adopt the latest foreign bridge designs to link Buda and Pest across the Danube. As well as acting as political gestures the schemes shaped Hungary's emerging banking system.

The early years of the British development corporations, set up in the early 1990s to encourage inner-city regeneration, were marred by poor design. The London Docklands Development Corporation sought a corrective through its infrastructure requirement for a series of bridges linking redundant quays. The relaxation of planning control in the area had resulted in a mixed bag of building stock but the bridge investment had the potential to be of exemplary quality. Through skilled manoeuvring the transportation manager, civil engineer Cynthia Grant, assembled teams of well-known designers to compete for the commissions. The third of six competitions was won by Future Systems, with architects Jan Kaplicky and Amanda Levete supported by the engineer Anthony Hunt Associates. Their scheme shows all the idiosyncrasies of a strong design philosophy. Kaplicky is renowned for his single-minded pursuit of form derived from modern technologies, while Hunt has contributed to many major examples of English high-tech architecture: the Sainsbury Centre, the Inmos factory and Waterloo Terminal. The many monographs on Future Systems reflect the polemical nature of their work and the West India Quay bridge demonstrates this quality throughout. It is a testbed for ideas, in which problems are carefully selected and their solutions identified.

The use of prefabrication was intensive. By adopting a pontoon configuration the entire structure could be made at the workshop and then transported between a bogie and a tractor unit to a dock-edge site for fitting out (the longest load to travel the motorways of England), before being towed up the Thames and moored in its final location. The end detailing, hinged plates just resting on the granite paving blocks of the quay, emphasizes the self-contained nature of the installation. The pedestrian steps onto this bridge across a safe but clearly defined threshold.

Permanent pontoon bridges are rare, partly because of their instability under applied loads; they bob up and down as people walk on them. The problem is resolved at West India Quay by tying down the bridge to tension piles in the dock bottom. Loads on the deck relieve the tie forces without driving down the floats. This device is structurally inefficient – the bridge could well stand on shorter compression units – but has secondary benefits. Levels and prestresses can be manipulated by pumping the pontoons, which is easier than jacking them, and the variations in the dock's water level are more readily accommodated. The floats also preclude a key problem associated with lightweight footbridges of this scale: pedestrian-induced vibration. The floats themselves, the buffered moorings to the piles and their embedded ends together add up to an extraordinary source of damping.

The structural hierarchy of the bridge is straightforward. The deck comprises planks extruded from aluminium (which needs no special corrosion protection because of its reactive nature). The strained-wire balustrade draws on the typical language of marine detailing. The deck

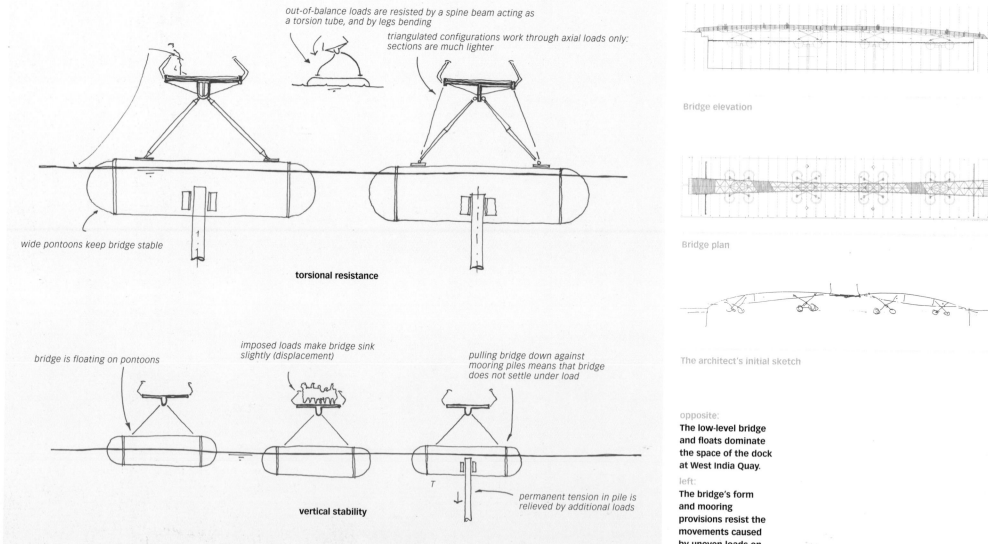

out-of-balance loads are resisted by a spine beam acting as a torsion tube, and by legs bending

triangulated configurations work through axial loads only: sections are much lighter

wide pontoons keep bridge stable

torsional resistance

bridge is floating on pontoons

imposed loads make bridge sink slightly (displacement)

pulling bridge down against mooring piles means that bridge does not settle under load

permanent tension in pile is relieved by additional loads

vertical stability

Bridge elevation

Bridge plan

The architect's initial sketch

and balustrade are carried by shaped ribs from a central spine beam that is fabricated into an elliptical section by splitting a rolled tube and welding in spacer plates. The beam, in turn, sits on crossed struts carried on the pontoon tanks, which are simple cylinders finished with pressed boiler ends. The configuration of the supports is forced. The clarity of the load path, from deck to ribs to spine, and then down to the legs and out into the floats, belies the problems created by the arrangement. Spectators standing along the deck edge create enormous twists in the central beam. The engineer's requests to put balancing cables to the outer deck edges were turned down and instead the structural weight had to be doubled.

The requirement for an opening central section was dealt with in the most obvious and direct way. This is perhaps disappointing given the potential for a floating structure to reconfigure itself temporarily. On the West India Quay Bridge two centre sections pivot upwards as in London's Tower Bridge. The counterbalances are detailed using familiar aviation references. The deck detailing is crisp, with neat runway lights along each side and concealed striplights in each rail emphasizing the long, low curve. The junctions within the steel framing are less well resolved: the connection between strut and float uses the most basic type of flange connection.

Parallels drawn between the bridge form and surface-skating insects are not accepted by the designers and yet there is a clear feeling of the deck skittering weightlessly across the water in a single curve, while the floats record the trajectory of a skipping stone. The almost naive but beautifully drawn profile of the walkway opening out to each end is strongly reminiscent of the Charles Bridge in Kaplicky's

home town of Prague, an ideal place to see how bridges work as sites for taking in the urban scene.

The critic Martin Pawley also cites the influence on Kaplicky of seeing the Americans arriving in Prague at the end of the Second World War in their unpainted silver Dakotas and B17s. Riveted monocoques recur in his work, and a preoccupation with flight can be identified in the earlier, more streamlined versions of the bridge's floats. Another influence might be the celebrated Schneider trophy planes from the heyday of heroic engineering.

A relentless aversion to compromise is evident in two features of the design. An initial sketch contains all the major features and must have been a continual reference as the design progressed. The pencil line representing the curve of the bridge is slightly distorted in the movement of the hand. That deviation from profile is religiously reproduced in the actual structure. A further disinclination to compromise is shown in the choice of colour, which is startling. It is a non-standard shade only seen on early models of BMW 3 Series cars, themselves a design icon. The colour was mixed and subjected to

expensive accelerated weathering tests before being used on the main frame. It is now marketed under the name of one of Future Systems' partners, Amanda.

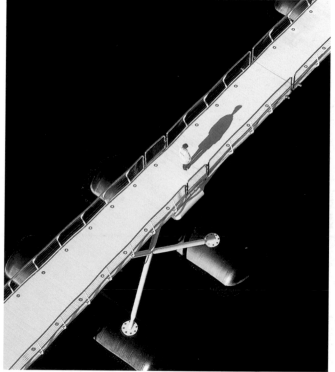

opposite:
Prefabricated sections were fitted out and commissioned in another redundant dock and towed up the River Thames to the site.

left:
The pontoons are pairs of simple cylindrical tanks with standard pressed steel boiler ends.

above:
The movement of the floats damps out pedestrian induced vibration.

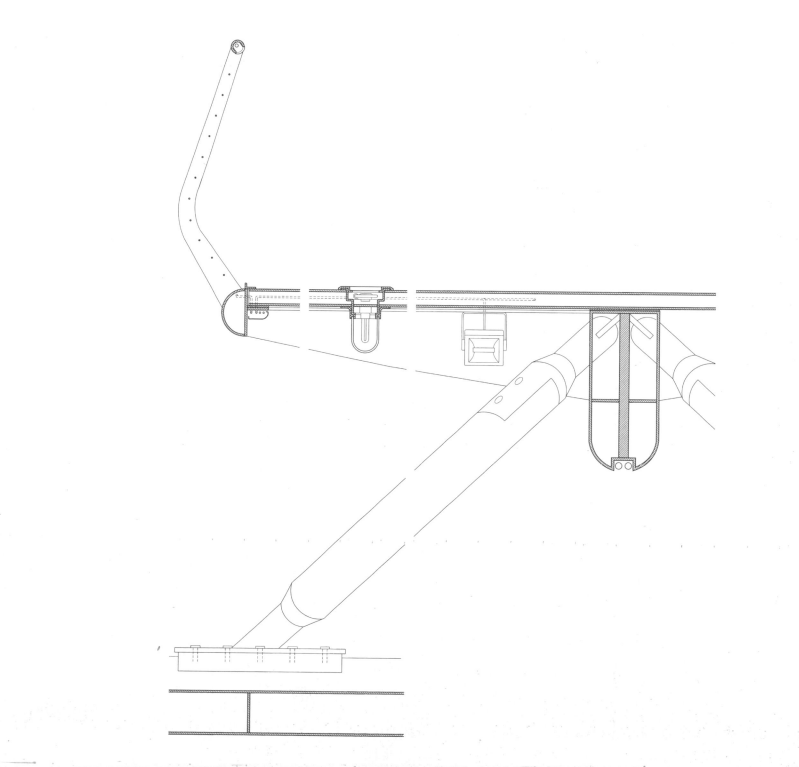

opposite:
The low curving profile of the bridge deck, funnelled plan and raked handrails result in a relaxed walkway.

left:
A central spine beam fabricated from plate steel carries outriggers and extruded aluminium deck plates. Runway lights set in the deck illuminate the balustrade. Handrail lights in turn light the footpath.

Kiel-Hörn Folding Bridge; Kiel, Germany; von Gerkan Marg and Partner / Schlaich Bergermann and Partner 1997
26.65 m / 87 ft

The ancient trading cities on Germany's northern coasts are characterized by brick merchant houses built alongside waterways, docks and ship-building wharves. Crossings are made with simple bascule bridges and other lifting mechanisms, which have evolved into a standard form that includes a low-level lightweight deck, overhead balance beams and counterweights. Kiel, on the Baltic coast, is split by the Hörn river into the old town to the west and the

shipyards along the eastern shore, which also takes in the less prosperous suburb of Gaarden. In the early 1990s the development potential of Gaarden and the pressure to improve communications became key issues in local mayoral politics.

The Scandinavian ferry company working out of the port agreed to move its terminal away from the river-front development on the condition that a bridge was installed linking the train station to the

displaced facilities on the far side of the Hörn. The structure had to be readily retractable to allow for a large number of boats, including local ferries and yachts, to move in and out of the port. Working to these constraints the chief planning officer, Otto Flagge, commissioned an extraordinary design.

His proposal was for a three-part bridge, developed from the two-element lifting-bridge format. It provided the opportunity to test out

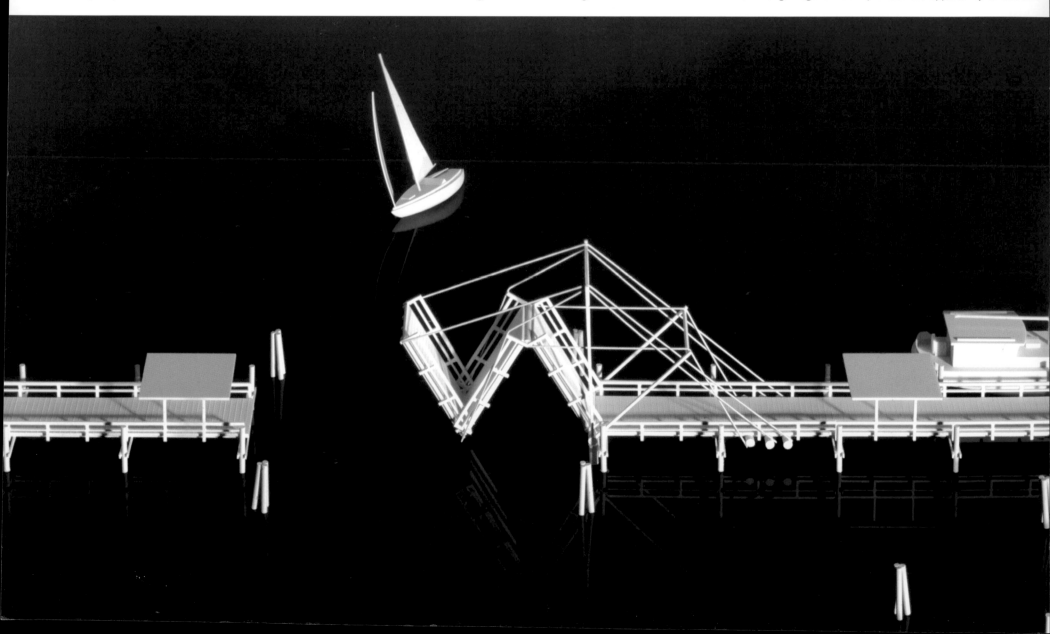

ideas and to create a symbol specific to the area. The result could be seen as a celebration of mechanisms: a simple concertina of deck panels, configured like a folding bed, is embellished with a cable support system stretched over two hinged frames; the sections are lifted and retracted by winching the stays.

The system has to function safely in the worst conditions of rain, ice and wind. Instead of employing sophisticated hydraulic or electromechanical systems the bridge uses block and tackle, with wire cables winding onto standard winches. The leverage gained through sheer leg frames and pulley systems is one of the oldest forms of mechanical advantage. This system was developed in an unbroken line from the early types of running rigging created by Hanseatic sailors for sail management to the heavy tackle of modern industrial fishing.

Schlaich's bridge mechanism centres on a single pair of continuous cables that can be steadily drawn in, and so avoids the problem of simultaneously retracting multiple leads. The stretch in cables cannot be precisely controlled in inconsistent conditions and they begin to pull unevenly, whereas a single cable moves smoothly through the system. This lifting mechanism remains stable at any point if its two and a half minute opening and closing cycle is interrupted.

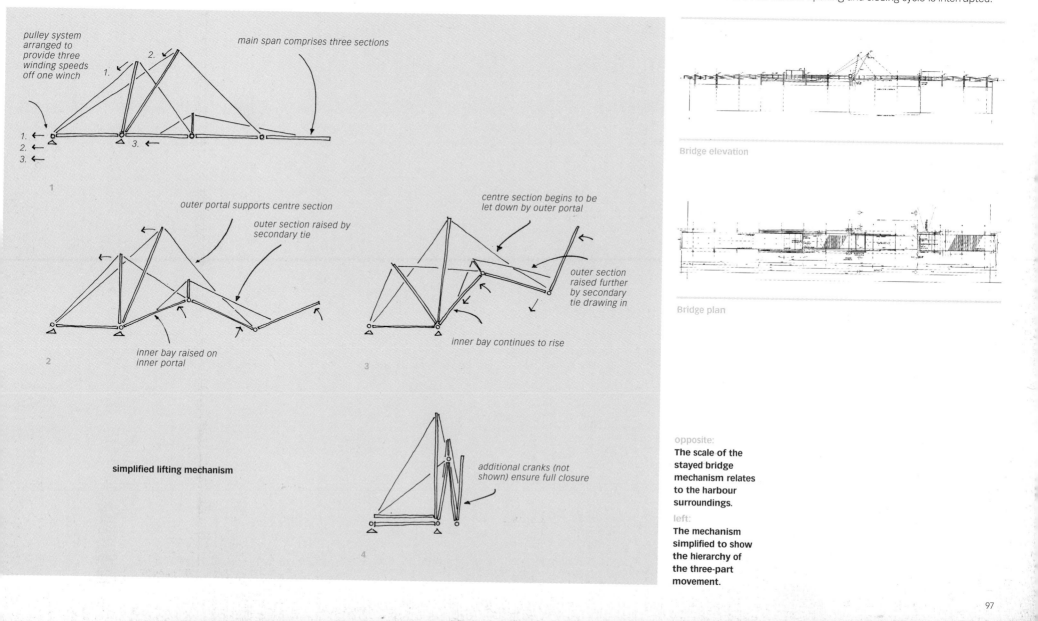

pulley system arranged to provide three winding speeds off one winch

main span comprises three sections

1

outer portal supports centre section

outer section raised by secondary tie

centre section begins to be let down by outer portal

outer section raised further by secondary tie drawing in

inner bay raised on inner portal

2

inner bay continues to rise

3

simplified lifting mechanism

additional cranks (not shown) ensure full closure

4

Bridge elevation

Bridge plan

The run of cables is displayed on the elevation of the bridge, so that the eye takes in the potential for movement. The transformation of stays from standing rigging (the support system) to running rigging (the mechanism itself) gives the structure a special wit: opening as many as 12 times a day, the bridge becomes a large kinaesthetic sculpture. Bright colours are used to draw attention to the articulation of parts, and to improve safety without resorting to strident markings to warn passing ships. Detailing on the structure is very understated and consciously avoids competing with the expressed complexity of the lifting gear. The timber deck matches expectations of a standard harbour installation's and the handrails are traditional posts and rails that fold neatly with the bridge's movement. Simple canopies set on the approach causeways provide shelter for pedestrians and cyclists waiting for the bridge to reopen. The shelter at the west end also protects the main winding gear set below the deck platform. The structure has a human scale in keeping with the passing vessels and nearby houses, and blends unobtrusively with its surroundings rather than drawing attention to itself as a design statement. The project itself, however, was very high profile, and its implementation was fraught with problems. Not only was there opposition from local politicians and inhabitants regarding the use of public funds, but when

the system was tested at the shipyard it was discovered to be too heavy to work.

The Kiel-Hörn Bridge illustrates the experimental approach taken by many of today's leading design engineers, but forays into areas that have not been tried and tested inevitably lead to unforeseen problems and to a necessary process of refinement. The bridge's mistakes were methodically diagnosed and rectified, proving the

design to be readily adaptable, but at the same time a media campaign and the rhetoric of the local elections raised speculation about the structure's condition and likelihood of success.

Opposition centred around the importance of funding other social needs in the area. Although this fact could not be disputed, it could be countered by arguing that the installation will have a lifespan of at least 120 years. An off-the-shelf substitute was temporarily installed

but the original design – 30 per cent over budget – was completed and erected three months later. Beyond the issue of the bridge's contribution to the region is, perhaps, the more problematic question of whether the cheapest utility can ever incorporate a real aesthetic.

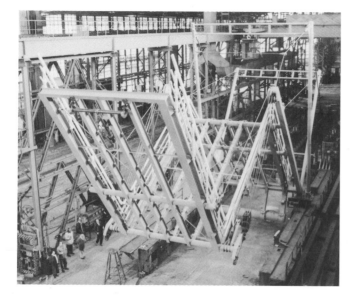

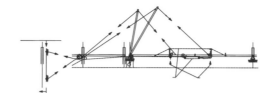

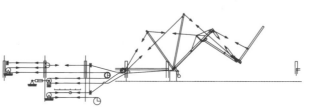

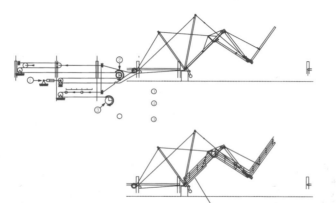

opposite:
The partially opened bridge against the background of Kiel's old town.

far left, top:
The main frame during trials at the assembly yard in Rostock.

far left, bottom:
Moving parts are emphasized as decoration and colour-coded to make the system's action readable.

left:
The several actions of the bridge are driven by one winch through a series of pulleys. The structure remains stable at all stages of its opening and closing cycle.

Fred Hartmann Bridge; Baytown, Texas, USA; URS Southern Corporation/Leonhardt, Andrä and Partner 1995
361m/1184ft

The unique appearance of the Fred Hartmann Bridge springs from two sources: its hurricane-prone site on the Gulf of Mexico and the pragmatic approach to construction in North America, where demonstrable efficiency and economy are key. The package is a sophisticated piece of engineering, in which careful thought has permitted the use of 'simpler' materials.

The bridge, which replaced a congested road-tunnel link, required eight lanes for traffic. Its exceptionally wide deck is potentially subject to very heavy winds and buffeting. The most dangerous dynamic

effects in bridges involve the deck edge lifting and twisting, and keeping the deck level at all times was therefore the chief safety requirement. Instead of building up this huge plate the designers kept it very thin in the interests of overall economy, but combined it with support towers and a stay arrangement of exceptional torsional stiffness. The A-frame end towers are combined with fan-stay groups to create giant 'torsion boxes'. The components are spread out in space so that the whole assembly – deck, towers and cables – is twisted by the wind, not the deck alone.

The designers examined several versions of the deck before establishing that girders separated into two parallel planes suited the system's aerodynamics. As the incidence and speed of wind changes, an object such as a bridge girder can rapidly transform from an aerofoil into a type of giant paddle damper. By separating the girders in this way it was found that at critical wind speeds, when one deck was fluttering in the wind, the lee-side roadway was shedding energy and calming the system. The double-diamond pattern of the towers is very efficient at resisting lateral pressure from wind. All forces are

resolved axially down the legs without bending them and, therefore the towers are significantly lighter than traditional H-frames.

The overall span of the bridge is within the range where simple cable-stayed configurations are efficient. The soil profile was uniform across the site and therefore did not influence foundation locations. One support was sited on an existing embankment built against floods and the other in the waterway, to give an ideal span. All the initial designs therefore assumed a completely symmetrical cable-stayed configuration, with the stays arranged in the conventional, statically

efficient, fan arrangement, which allowed for uncongested anchorages on the towers. Uneven loads in the centre spans are balanced by a group of three cables at the back of each fan fixed down to the end piers. These backstays tend to compromise the fans' clarity and simplicity. All the stays were sized so that any one of them could be damaged in service without danger. The worst catastrophe envisaged was a tanker from the nearby oil town catching fire on the bridge, so the backstays were given a substantial cement-grout fire jacket over their lower ends. Again, these compromised the bridge's visual

balance. Various multiple-stay systems were considered but rejected on economic grounds because the cost of the extra cables and their anchorages did not justify the consequent weight reduction in the deck. Ease of construction and economy of materials were central to the design process. Four schemes, each using different materials, were developed by the consultant teams and then tendered. Extensive redesign and development by the contractor then took place before arriving at the built project. Of the optional materials, the universal choice among those tendering was a composite deck

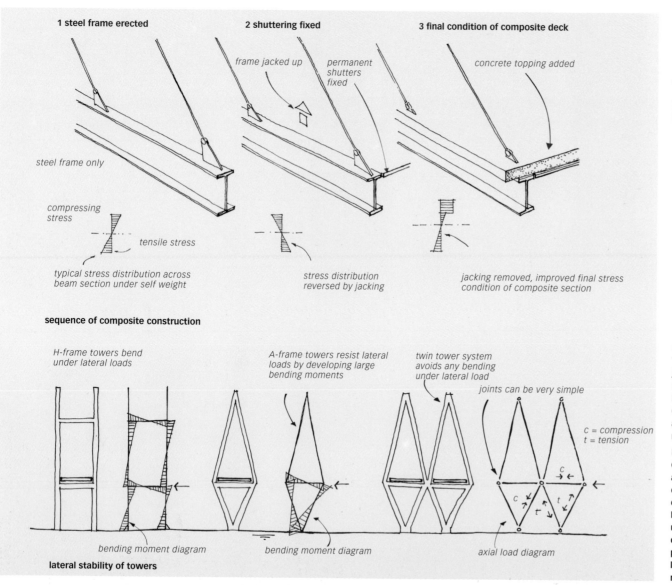

1 steel frame erected

steel frame only

compressing stress

tensile stress

typical stress distribution across beam section under self weight

2 shuttering fixed

frame jacked up

permanent shutters fixed

stress distribution reversed by jacking

3 final condition of composite deck

concrete topping added

jacking removed, improved final stress condition of composite section

sequence of composite construction

H-frame towers bend under lateral loads

bending moment diagram

A-frame towers resist lateral loads by developing large bending moments

bending moment diagram

twin tower system avoids any bending under lateral load

joints can be very simple

c = compression
t = tension

axial load diagram

lateral stability of towers

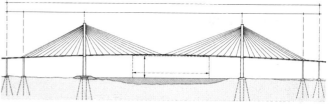

Bridge elevation

opposite:
The wide road bridge joins La Porte and Baytown over the Houston Ship Channel, creating an integrated oil refinery centre.

left:
Composite steel and concrete construction is employed for the bridge deck. The unique braced tower configuration provides exceptional lateral stability to resist hurricanes.

structure of steel and concrete with concrete towers. The subsequent development of the design centred around controlling movements within the structure and the plant requirements for different detailed configurations.

The deck is a composite element comprising a concrete road base supported on a grillage of steel beams. The advantages of combining these materials include a significant reduction in weight compared with an all-concrete structure and the opportunity to exploit their interaction. During construction the steel frame acts as a light deck,

supporting only itself and some access, and is sized accordingly. Precast concrete panels are fixed as permanent formwork, and the frame is then jacked up slightly. Concrete is poured over so that the whole assembly locks together. The jacking introduced a favourable distribution of internal stresses and the composite section works efficiently to support the final loading condition in a way not readily achievable with steel or concrete alone. The concrete prevents the steel from buckling locally and the plate sections can therefore be very thin, reducing welding and heat distortion problems.

However, over time and under the pressure of load concrete changes its shape. The material forms by absorbing water in a chemical reaction that hardens the pourable paste into a stone-like substance. This process continues over time, causing shrinkage. Furthermore, concrete acts plastically when placed under load, yielding very gradually to the applied force.

At this scale such movements become significant and are compounded by the effects of temperature. In its early years the bridge will shorten by 15cm (6in). To accommodate this movement the

deck is preset with a camber and allowed to arch slightly so that the concrete topping always remains in compression. During construction the outer piers were jacked apart before the deck was completed so that they will gradually return to an upright position as movement occurs.

The contractor's method followed the proven path of cable-stayed construction, in which the towers are erected and then the deck formed as a balanced cantilever creeping out simultaneously on both sides of the towers. Wind-tunnel tests showed that this would be a vulnerable time, and temporary tie-downs were therefore added. The resistance of the bridge to longitudinal loads was ensured by placing neoprene bearings between the bridge deck and the towers, which distribute stresses evenly to both sides. These forces and an allowance for imbalances during construction make the towers much wider at the sides than at the ends viewed from the road. Lack of work space resulted in the construction of a substantial artificial island around the west tower, which has since been used as ship-impact protection. The contractor made major alterations to the design to suit the construction method adopted. Prefabricated deck elements were lifted up off barges by a crane that was standing on the structure rather than floating in the canal. The weight of each component had to be kept down and so the structure was built up gradually. The steel framing was fixed and then the precast concrete panels were lifted on and grouted into place like giant tiles. The bridge superstructure weight rose significantly because of these changes but the overall price went down. The final joining up of the end spans took place at night so that temperature changes did not compromise accuracy.

Steg über der Mur; Graz, Austria; Günther Domenig / Harald Egger, 1992
55.8 m / 183 ft

The city of Graz has defined itself through its architecture. The regional capital of Steiermark, and the second city of Austria, it is cut off from nearby Vienna by mountains and turns towards Hungary and the Balkans, forming a crossroads of south-east Europe. Its border location and secondary status encourage outward-looking sensibilities. International modernist ideas overlay a subdued, critical regionalism. The surrounding region has the best reserves of coal and iron in Austria and, because of late industrialization, maintains an expertise

in steel, glass and aluminium handling: a palette of materials. In the 1970s Christian Democrat politicians set up a party initiative entitled the Modell Steiermark, defining the nature of public building. In Graz a leading protagonist of the movement, Josef Krainer, was appointed the head of Graz's architecture department in 1976. He took the radical step of commissioning all new work from independent architects. The continual stream of competitions, exhibitions and published works kept public consciousness at the highest pitch of critical awareness.

The products of the Modell Steiermark became known as the Grazer Schule, although the label is rejected by some of those involved who do not recognize a coherence in this work. There is, however, a discernible spirit in which projects are carried out. For instance, the architecture critic Peter Blundel-Jones suggests that a preoccupation with bridges, thresholds and edges as transitional spaces is characteristic of the new Graz architecture. In 1990 a competition to design a footbridge over the Mur was held among architects

indigenous to the Steiermark. A scheme by Günther Domineg, the dominant figure in Graz's architectural scene, won the prize from among 36 submissions. Domineg was a student and later a professor of the city's school of architecture. The school is one of only three in Austria and has a formal style of teaching. Starting out in 1963 Domineg explored Brutalism, with its unrefined materials and direct detailing, before moving on to an unrestrained expressionism. His footbridge links the old town and the Mur suburb. Two potential sites

were identified in the competition brief, and Domenig chose the direct line between the Marienhilfekirche on the suburb side and the approach to the old town's castle.

The bridge is laid out as a central plate box beam, triangular in section and joined across a central pin junction. A central strut bears onto a bow-string tie made up of four parallel tie plates. The primitiveness of the system – it uses the smallest number of components that could together be styled a truss – gives it a solidity. It appears to be on a

disproportionately large scale for the simplicity of its form. The design process can be read as starting with the treatment of the bridge as a pure type: the truss as a diagram of forces that is then inflected by the vagaries of site and section.

The change in levels between the banks (2.2m/7ft) is dealt with subtly. The deck rises at the maximum allowable pitch to the centre break and then inclines very gently to the far side. The supporting girder rises through the ramped section but remains below the upper

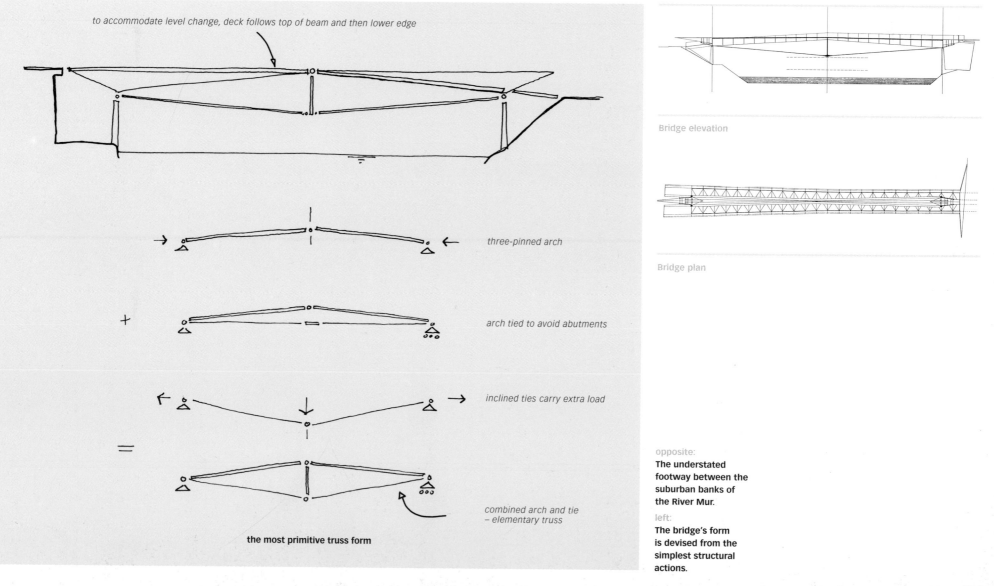

to accommodate level change, deck follows top of beam and then lower edge

three-pinned arch

arch tied to avoid abutments

inclined ties carry extra load

combined arch and tie – elementary truss

the most primitive truss form

Bridge elevation

Bridge plan

105

flat part so that the beam soffit and tie assemblies form a level and symmetrical composition above the water. The walkway comprises timber decking and steel footplates supported on cantilever brackets. It is a secondary element to the spine beam, responding to the idiosyncrasies of the site, while the structure remains self-contained.

The site is on the bend of the river. The erosion face (the outer side of the curve) is revetted with a vertical retaining wall, while the other bank slopes gently down to a riverside walk. The bridge stands on

simple paired stilts on each bank. These piers serve to detach the structure from its surroundings both visually and literally, in separating it from obstructions in the old embankments. The overhanging bridge ends touch each side lightly on slide plates. On the south side the bridge divides into two walkways as it descends on each side of the central spine beam. A staircase leads down through the gap to a viewing platform on the sloping bank. This observation point for viewing the soffit reflects the conscious self-referencing of the design.

The sharp finial above the stair is justified by the architect as a marker in the townscape.

Detailing on the footbridge is idiosyncratic. The lower-boom tie plates are gathered together onto pin ends. These are expressed in a plate assembly at the lower centre connection and then suppressed within the box sections above. The substance of the centre prop is reduced by composing it as a cruciform assembly of plates. Light secondary stays have been retrofitted to prevent the tie rods, which

have no sideways stiffness, from shaking. The pin-ended cables are, unfortunately, an intrusion. Balustrades are formed from clear panels of toughened glass in stainless steel frames, and a tubular top rail houses lights that shine onto the walkway at night. Lighting directed onto the structure emphasizes the soffit and suppresses the visual effect of the asymmetries.

The bridge flexes and sways lightly under foot, a function of the simple plane-frame arrangement of the structure and the rising profile of the deck. As a result of the layout applied loads tend to destabilize the structure, and only the central tubular-steel section resists torsion, twisting forces about the long axis. The concentration of resistance along the bridge centre line is inefficient. The weight of the bridge – 117 tonnes (115 tons) spread over 53m (174ft) – is more than 2 tonnes per metre (almost 1500lb per foot) and is twice the usual benchmark, indicating the ascendancy of architectural over engineering considerations.

This combined footbridge and cycle way plays an important role at Graz, as indicated by a local commentator, who describes it as a 'place for humanity, a symbol and a cultural axis for the city, where a tension field between tradition and receptiveness to the new has been set up.'

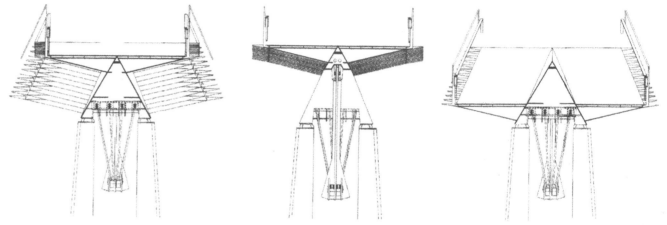

opposite:
A sculpted marker indicates the stairs down to an observation platform on the west bank. The old town castle of Graz is in the background.

left:
The tie rods and central prop are built up from flat steel plates.

above:
Sections taken along the bridge show the angular development of all its parts.

Allmandring Footbridge; Stuttgart, Germany; Kaag and Schwarz/Gustl Lachenmann 1994

34m/96ft

Small bridges offer an opportunity to experiment, to explore the ramifications of a structural mechanism or an applied idea through set pieces. In contrast, large projects inevitably summon up a conscious conservatism. There may be few risks attached to a large project, but the consequences of failure make the expectation of success disproportionately high.

Allmandring Footbridge provides an inexpensive and expedient solution for a university campus. Stuttgart University's role in the development of lightweight structures is well documented, particularly its contribution to membrane structures and to small- and medium-scale cable handling. The footbridge reflects this expertise.

This small *Steg*, or gangway, was conceived as an investigation into structural action. Instead of making a series of controlled experiments in which one variable is systematically altered, a complete system was set up and then closely observed. The scheme, which is the result of a local competition, embodies a new structural mechanism. The competition was presided over by the engineer Jorg Schlaich, a contributor to the Stuttgart Lightweight Structures Laboratory and well known for his engagement with environmental issues.

The footbridge employs a simple arch of deck plates, articulated into eleven hinged sections. This light ring has almost unlimited strength under even loading but quickly distorts and then buckles under variable pressure. A second system is therefore applied to the first: a horizontal pre-tensioned bar attached to each abutment is

connected to the deck with a rod and acts as a stabilizing element. When a load is applied to one side of the bridge it twists the deck and forces the tie rod sideways, but this deflection sets up compensating tension forces that pull the system back into shape. The structure becomes an exercise in balancing out disturbing forces with the stabilizing force pattern set up by the central tie.

Since this heavy rod is stressed against the abutments no load is transferred to the arching deck. The deck is therefore extremely light,

a quality that is accentuated by its surface, an open grid which is firm, self-draining and always frost-free. Temperature changes reduce the effectiveness of prestressed elements, particularly in metal structures, but this small bridge is just below the scale at which losses and other secondary effects become significant.

The abutments are shaped blocks of concrete, simple elements that effectively combine several functions. Acting as a clearly defined threshold between the ground surface and the deck, they house

spacious manholes in which the main rod ends can be accessed. The realignment of forces from the main rod into the ground anchors is smoothly accomplished. The tensioning of the main bridge against each abutment, which produces large lateral forces that are sustained by ground anchors, is standard technology for membrane tie-downs.

Lateral loads from the balustrades, wind loads and the sideways forces transmitted by walkers are resisted by cross-wires set into the deck frames. These braces are prestressed to enhance their action.

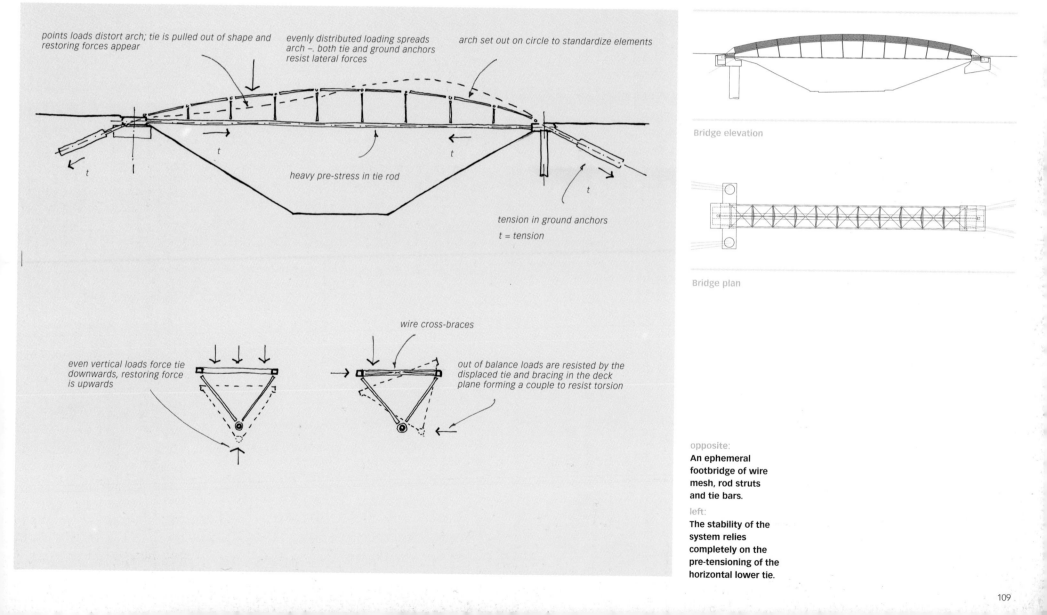

points loads distort arch; tie is pulled out of shape and restoring forces appear

evenly distributed loading spreads arch –. both tie and ground anchors resist lateral forces

arch set out on circle to standardize elements

heavy pre-stress in tie rod

t

t

t

t

tension in ground anchors

t = tension

Bridge elevation

Bridge plan

wire cross-braces

even vertical loads force tie downwards, restoring force is upwards

out of balance loads are resisted by the displaced tie and bracing in the deck plane forming a couple to resist torsion

opposite:
An ephemeral footbridge of wire mesh, rod struts and tie bars.

left:
The stability of the system relies completely on the pre-tensioning of the horizontal lower tie.

Tensioning both cables within the rectangular frames holding each deck grid guarantees that the structure will be undisturbed: when the frames are distorted, tension increases in one cable but diminishes in the other. This reduction relieves the system, which then appears to react less to the imposed load. The flexible cable behaves like a rigid strut.

The structure undergoes extensive distortion in order to resist applied loads, and the consequences of accommodating this movement run throughout the detailing. The various parts must be fully separated by joints to mitigate the effects of uneven pressure. In light truss bridges secondary bending forces can lead to an increase in the size of many of the elements. The removal of these forces by using pins or spigots on the end of each strut and cable ensures that this bridge requires minimal section sizes. Pins are a long-term maintenance problem, however, and would be a significant cost on a larger structure. Detailing is deliberately straightforward, with standard socket ends and industrial gratings, but its rigorous execution raises it above the merely prosaic. The balustrades are gauze-like panels of chain-link fencing supported on stretched cable rails. Forged and shaped fork ends, cover plates, cable sockets and turn-buckles add interest to the composition.

The tautness of the whole system is expressed in the proportions of its elements: the posts are of a similar size to the edge rails, the edge rails are of a similar girth to the tie bar. The even spread of parts gives the overall impression of forces held in balance.

The main tie rod uses a sample left over from testing the ring-tie system for a large sports hall roof. It is stainless steel, and the same specification was used in the adjacent struts so that connections could be made directly, preventing the formation of corrosion crevices and paint traps, and visually expressing the load paths. As a further economic measure the decks are made from galvanized mild steel. This material had to be isolated from the underlying stainless steel to avoid the problem of bimetallic corrosion. (The two metals have very different valencies and combine to set up a galvanic cell – a battery – in which the stainless steel rapidly corrodes.) The neoprene isolating layer employed in the footbridge simultaneously acts as a damper, attenuating the impact of footfalls on the deck panels as they pass through to the underlying frame.

The structure has proved to be very flexible. The natural frequency of sideways vibration is within the range of normal human motion, and the structure sways under pedestrian loading. Analysis shows that dampers at quarter points on the span would be effective in drawing off the accumulating energy, but these have not been fitted. The bridge's designers consider its response to be a useful part of the students' experience of the structure. It has become a didactic tool: students in the engineering school are expected to observe the bridge's exaggerated action and to consider its implications. The process may reveal new ways of controlling pedestrian excitations.

opposite, left:
The bridge deck is set back inside the edge beams so that the balustrade can be buttressed with raked struts.

opposite, right:
Lateral loads are resisted by a cross-bracing of fine wire strainers beneath the deck plane.

left:
The cable anchorages are integrated into the corner stiffening plates. Walkway and balustrades are of the simplest weld mesh and chain-link fencing.

Royal Victoria Dock Bridge; London, UK; Lifschutz Davidson Design/Techniker 1998
127.5 m / 418 ft

Silvertown, which forms part of London's Docklands, is isolated on a narrow stretch of land between the vast Royal Docks complex to the north and the River Thames to the south. Notoriously a depressed area, but adjacent to City Airport and labelled as one of the 'gateways to London', it became a focus for regeneration in the late 1980s. A large parcel of the land was sold to a private housing initiative to create an urban village of unprecedented size. As part of the deal an agreement was made with the development corporation to open up access to the north quay. The line of the new structure would also continue the linear route taken by a local park.

A design competition was held for the crossing. The brief called for a covered way raised high enough to allow boats belonging to the sailing club housed in the dock to pass beneath. At the same time the overall height of the structure was limited by the glide paths of the nearby airport. The winning team, led by architects Lifschutz Davidson, avoided the prospect of a dirty and dangerous elevated tunnel by proposing the solution of a manned transporter car running below an open deck.

Transporter bridges were patented by the French engineer François Arnouldin, a pupil of Alexandre Eiffel. Only 22 examples were built, in the late nineteenth and early twentieth centuries, before the system became obsolete. At the Royal Victoria Dock this system could provide the air draught necessary for sailing vessels and offer a commercial advantage. Imposed loads are automatically limited by

the size of the car, which means that superstructure weights are correspondingly reduced, resulting in both the classic attenuated silhouette of this invention and a low initial cost. Although at first the competition judges favoured more elaborate submissions, eventually the least expensive solution was chosen.

A wide range of influences inform the design. Lifschutz Davidson always complete their work with a careful clarity of detail, reflecting their apprenticeship on Foster and Partners' high-tech structures of the 1980s. Large-scale gadgets often feature in their projects. In this case the car system and the exposed lift mechanisms on the access towers are an intrinsic part of the design's expression. The faceted deck profile is lit by a beautiful system of point sources incorporated like pocket-knife blades into the cast uprights of the balustrades.

The car trolley is supported and driven on a single set of centrally placed wheels to avoid 'crabbing' (the term used when spaced wheels get ahead of one another and twist up). Sideways swing is prevented by incorporating stabilizing wheels that press upwards against the soffit. The main bridge beams, which were sized to the limits of road transportation, are arranged centrally to carry the running rails, and the deck is visually suppressed so that it looks like a secondary add-on. The beams are drawn up through the deck to minimize wind resistance; their humped tops break through the walking plane as an insistent reminder that structure is taking precedence over service-ability. The box girder shapes are structurally efficient lenticular

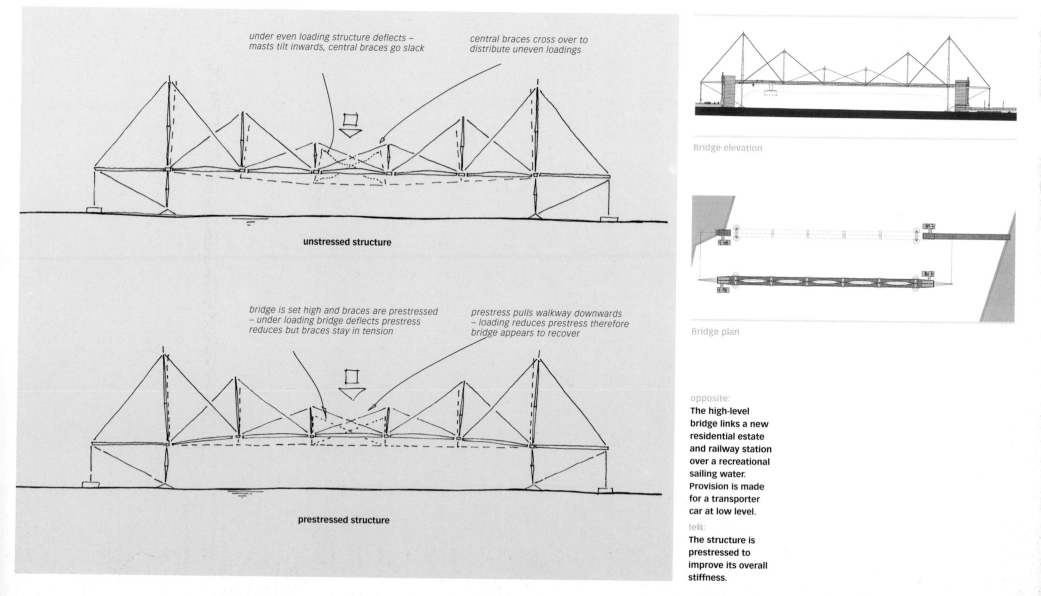

under even loading structure deflects – masts tilt inwards, central braces go slack

central braces cross over to distribute uneven loadings

unstressed structure

bridge is set high and braces are prestressed – under loading bridge deflects prestress reduces but braces stay in tension

prestress pulls walkway downwards – loading reduces prestress therefore bridge appears to recover

prestressed structure

Bridge elevation

Bridge plan

opposite:
The high-level bridge links a new residential estate and railway station over a recreational sailing water. Provision is made for a transporter car at low level.

left:
The structure is prestressed to improve its overall stiffness.

113

profiles modified to suit the requirements of the soffit running gear and then simplified to ease fabrication. Great care had to be taken in the sequence of welding to avoid gross distortion of these odd shapes.

A wind tunnel became one of the design tools in the scheme's development. Environmental tests were made on models of the bridge to show that the structure would not significantly alter the wind conditions in the dock. In addition, the balustrades were adjusted to direct air movements, in order to shelter users as much as possible. The proposed design was extremely light and flexible. The engineers

took a conscious decision to confront aerodynamic effects rather than to shy away from them, and therefore chose a deck section that is capable of acting aerodynamically. A test model without a balustrade began to oscillate at low wind speed; it developed a vertical galloping motion combined with a twisting action. This is known as 'classic flutter' and is the most dangerous of all aero-elastic instabilities. The addition of a solid balustrade again caused vertical galloping, within the range of foreseeable conditions. The best option was a barrier perforated by 35 per cent, which remained stable in the

strongest anticipated winds. Air is forced through the sieves of metal, which become aerodynamic dampers, shedding energy.

Many options were reviewed in the effort to find a simply expressed cable configuration. The primitive lines of the Fink truss were eventually adopted in a diminishing sequence of simple sub-assemblies cantilevered to interlock across the centre. The arrangement reveals both its mechanism and its means of construction. The structure as a whole is given apparent stiffness by being prestressed across the centre section to pull it down into its final line. When the bridge

dips under a load, the prestress is relieved and the bridge then rises in compensation.

The structure's maritime character stems not only from its details but also from the implicit idea that it might have been assembled from discarded materials. The Royal Victoria Dock was once the destination of enormous steel clippers that plied between Britain and Australia carrying grain. Spare sets of spars were kept by dockyard riggers to repair storm damage, and a bridge could have been constructed from such parts.

The original embankments of the dock were built over with reinforced-concrete wharf platforms. The backfilling contains many large obstructions, and therefore the main piers were formed by piling inside the dock. The deep, dense gravel beds on which London's skyscrapers are founded provided a firm (and economic) footing below the water. Unfortunately the possibility of finding unexploded Second World War bombs in the dock had not been anticipated and this complicated the construction process. The superstructure was erected off jack-up barges carrying scaffold towers as temporary supports. The very large lifting capacity of floating cranes was exploited by prefabricating the deck sections, balustrades, lighting and trolley mechanisms at the dockside before lifting the fully assembled units into position. The structure was given a transient quality by being fully demountable: it was pinned and pulled together; nothing was bolted or welded on site.

The final design is very much a reaction to the modernist intrusions found elsewhere in the area. The architects have tapped into a real sense of place by matching the bridge to the scale of the docks.

opposite, left:
The main longitudinal beams are pulled up behind the balustrades to streamline the bridge deck.

opposite, right:
The walkway deck is cantilevered off a central spine which carries rails and power supply for the transporter trolley.

far left:
The entire structure is erected without bolts or welding.

left:
The bridge's cross-section showing the aerodynamic box beam and balustrade configuration developed in the wind tunnel.

Talmadge Memorial Bridge; Savannah, Georgia, USA; T.Y. Lin International 1990

337m / 1106ft

Studies of contemporary bridges often centre around the range of influences to which designers respond. In direct opposition to this the Talmadge Memorial Bridge sets aside all considerations other than the necessity to link two points in the safest and most economic way.

In keeping with this philosophy, the bridge has the minimum number of working parts. It takes the simplest of forms, the cable-stayed bridge, whose mechanism is even more obvious than that of a beam and post construction. (Stayed walkways, developed from the rigging and spars of sailing boats, are an ancient device.) Its form reflects its load and the way it was made, and contains nothing superfluous.

The bridge links Savannah with Hutchinson Island across the Savannah river, replacing an old cantilever truss that was too low and whose piers had been struck heavily three times. The new bridge required another 15.5m (50ft) of air draft and supports on land. The highly experienced bridge-engineering firm of T.Y. Lin took on the entire design of the new crossing and, apart from expressing a preference for a slender silhouette, concentrated on simplifying the components and maximizing value and construction safety.

Concrete, steel and composite options were examined, and an all-concrete scheme was found to be the least expensive. (This type of structure, made from concrete reinforced with rods, originated in

the need to overcome a scarcity of other materials and to reduce costs.) Two towers are set up on each bank, carried on deep, bored piles through the weak delta sediment. Drilling holes for piles brings up a precise record of the foundation soils so that each shaft does not have to be oversized to account for uncertainty. The alternative of driving piles to a set level of resistance to hammer blows is a less precise method.

The towers are simple rectangular sections built up into frames. Their lower elements are tapered to resist the high, lateral wind loads and the upper parts are parallel to minimize shuttering. The mid-height crossbar is fixed at the best position to enhance the tower's lateral stiffness and allows the two sides to incline slightly, passing around the road deck but still picking up the cable groups aligned vertically over the deck edge beams. The fans are set out to

allow plenty of room for the anchorage points, whose intersection geometry is the simplest possible.

The deck is a continuous concrete element across the bridge's three spans; the cross-section does not vary along its length. Two edge beams support spaced cross-beams and a simple reinforced-concrete slab roadway. The floor beams in the central and end-span sections of the slab are prestressed by different amounts to cope with

a large part of a suspension bridge's lateral stability comes from the 'pendulum effects'

suspension bridge

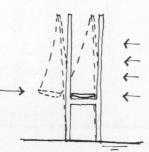

when the wind displaces the deck sideways the cables become inclined – part of their tension generates a restoring force

cable-stay bridge

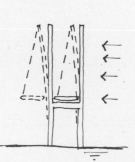

in a cable-stayed bridge there is no 'pendulum effect' – the displaced cables still just keep the deck in axial compression: a stronger deck than in a suspension bridge is required

the lateral stability of suspension and cable-stayed bridges

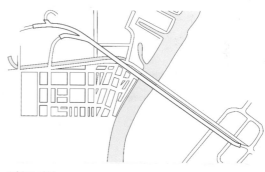

Bridge elevation

Bridge plan

opposite:

A cable-stayed bridge with no superfluities.

left:

The differing lateral resistances to load of cable-stayed and suspension bridges.

widely varying forces along the uniform girder's length. To allow for thermal and long-term concrete shrinkage and creep, each end of the bridge slides on a bearing at the junction with the approach ramp, which are both simple prestressed-concrete beams on hammer-head piers. Each half of the bridge balances on either side of its supporting tower. Unevenly distributed imposed loads in the roadway are counteracted by the upper backstays, which are bunched together and anchored into the transition pier.

A monolithic deck suits the cable-stayed bridge form. In a suspension bridge, lateral stability comes from the 'pendulum effect', in which lateral displacement of the suspension cable induces an opposite force. This does not happen on a cable-stayed structure, where resistance to lateral wind load is provided by the deck plate alone, which acts as a horizontal beam between the towers. As a result the roadway must be made much heavier. The mass of the deck contributes to the structure's dynamic stability, since its inertia

makes it less responsive to disturbing forces and keeps natural frequencies low. Concrete, which has been microscopically cracked by hydration shrinkage, also provides a high level of internal damping in the system.

Simplifying and reducing the number of parts made the bridge very easy to construct. First the towers were cast in stages with lifting forms. Then the preliminary sections of deck, the pier tables, were fixed between the legs and the spans constructed outwards from the

towers towards the centre and back to the approach ramps. The monolithic girder was cast in sections on movable formwork that could be projected alternately from both ends of the deck segments. The project engineer, Man-Chung Tang, had extensive experience of the process and used an improvement pioneered on a similar structure in Jacksonville, Florida: the traveller's weight is reduced by supporting the front edge on the permanent cables, so that a back section to balance the cantilever becomes unnecessary.

The design of the finished bridge owes little to its surroundings, referring to nothing extraneous. The vertical load-bearing system, the mechanisms to maintain lateral stability and the inert resistance to dynamic loads are all readable in the form. The rectilinear, parallel surfaces and arrised edges, the surface colouring and the formwork markings all reveal the construction method. Although the bridge has been reduced from an individual structure into a general type that is dissociated from its location and the specific reasons for its

construction, it can nevertheless be precisely placed: it is a study within an evolving body of work by the current, pragmatic generation of North American bridge engineers, and therefore is a freeze-frame in a continuum of ideas.

photos ©Scott Jolliff

opposite, left:
The main tower trestles are shaped to resist sideways forces by bending from the base upwards.

opposite, right:
Slight sags in the main stays are readily visible. Under live loading these contribute additional unwanted extension to the cables, which had to be allowed for in the design.

left:
The workman-like lighting emphasizes the bridge's industrial demeanour.

121

Charles River Mainline Bridge; Boston, Massachusetts, USA; Christian Menn / Bechtel Parsons Brinckerhoff 2001
227 m / 745 ft

Boston's 'big dig' is the largest single infrastructure construction project yet undertaken in a metropolitan area of North America. It involves a complex of tunnels, intersections and bridges, which have been built to replace the old Interstate Highway 93 and extend the Massachusetts Turnpike to the city's airport. The existing viaduct bisecting the downtown area has been removed, opening up the urban centre to its waterfront. The city is regularly gridlocked and the 14-year programme of works had to continue in the midst of this congestion while contributing to it as little as possible.

Initial designs included a conglomeration of ramps and bridge links defined by high-speed road geometries, fracturing the dense urbanism of Boston's North End and Charleston across the water. Universal dislike for the proposals led to an extended design study in the search for a unified bridge scheme that could resolve foundation and navigation problems, and provide a 'signature structure' for the multi-billion dollar project.

The interstate highway crosses the River Charles in North End at the same point as Paul Revere on his historic ride. The crossing is divided into two bridges. The Storrow Drive Connector Bridge, the largest steel box girder structure in the United States, carries four lanes of local traffic, and alongside it the Charles River Mainline Bridge supports ten lanes of the interstate highway. This accommodation creates a bridge of inordinate width.

In total 16 schemes were considered for a new crossing to replace the Mainline Bridge, including arches, trusses, box girders and several forms of cable-stayed structure. Two finalists were selected, both offering a cable-stayed bridge, one with a single pylon and the other supported by two main piers. The single support proved too tall for the urban setting, and therefore the twin-tower concept, prepared by the veteran Swiss consultant Christian Menn, was eventually adopted.

Menn's application of the classic fan cable-stayed bridge form is modified to suit the convoluted approach roads and complex foundation conditions. The forestays spread out to the edge girders but the backstays on the south side are gathered onto the bridge's centre line to avoid

any conflict with the Storrow Drive approaches directly below. The need to keep traffic flowing during construction directly influenced the design.

The central span is a steel and concrete composite deck. Box girder edge beams pick up shaped cross-beams on which precast-concrete deck panels are arranged, and the assembly is locked together with poured concrete closure strips. The steady build-up of smaller components was used to simplify construction in the confined conditions.

The bridge's back spans are prestressed-concrete box girders that were cast in place. They are heavy components capable of balancing the main span in the minimum of space. The cable spans support each side of the main deck and come down the centre line of the back spans to reach concentrated foundation positions spaced along a spine beam. This configuration generates an inverted Y-shaped pylon form, a central spire containing the anchor blocks springing from a steeply sloped A-frame straddling the roadway.

Below the roadway the frames rest on full-width pile caps. The patterns of large-diameter drilled shafts below are adjusted to suit the surrounding obstructions, services and tunnels that had to be retained. On both north and south sides the pylon foundations are sleeved in jackets, isolating them from the surrounding ground so that forces are not transferred into a nearby tunnel, which is used for the Orange Line Subway train, but pass down to the underlying bedrock. Transmitting loads into deeper ground strata limits the problem of draw down. When a heavy object is placed on the ground, the surroundings inevitably settle and nearby buildings may be affected. Deep footings disperse the distortion.

The cross-section of the bridge is asymmetrical. On the main span eight lanes are supported between the fans and two on a cantilevered side-piece bolted to the main deck. The eight-lane section alone forms the backspan, because of the necessity to minimize traffic disruption on the landsides during construction, and the two additional lanes are supported independently. Together with the bridge's exceptional width

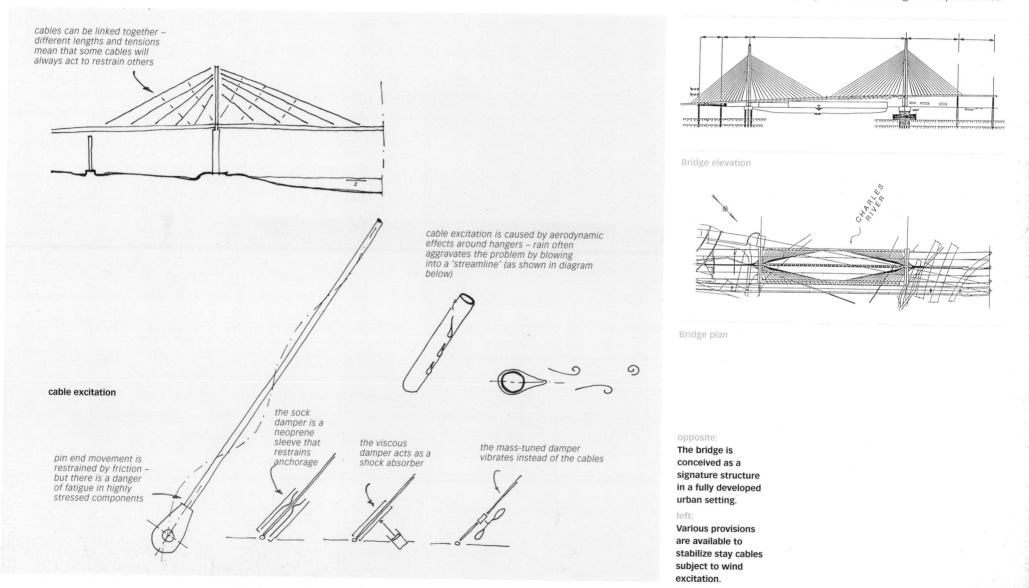

cables can be linked together – different lengths and tensions mean that some cables will always act to restrain others

Bridge elevation

cable excitation is caused by aerodynamic effects around hangers – rain often aggravates the problem by blowing into a 'streamline' (as shown in diagram below)

CHARLES RIVER

Bridge plan

cable excitation

pin end movement is restrained by friction – but there is a danger of fatigue in highly stressed components

the sock damper is a neoprene sleeve that restrains precast anchorage

the viscous damper acts as a shock absorber

the mass-tuned damper vibrates instead of the cables

opposite:
The bridge is conceived as a signature structure in a fully developed urban setting.

left:
Various provisions are available to stabilize stay cables subject to wind excitation.

this unusual arrangement created several problems that had never been encountered before.

The imbalance in weight between the two road sections was resolved by using lightweight concrete for the two-lane side section. To avoid problems caused by the absence of axial loads, and by the different shrinkage characteristics in this element, it was prestressed to experience the same 'load environment' as its neighbour and therefore moves in unison.

A symmetrical distribution of stays might have carried the out-of-balance loading from the side section, but the towers would have suffered serious twisting forces and the deck would have been pulled sideways. Concerns over long-term creep effects in the concrete, gradual distortion sideways and resulting changes in shape proved

too great. The upper cable anchorages were shifted to give the force pattern, although not the structure itself, a stable symmetry. Offsetting the line of the backstay cables produced the same effect and was shown to be a feasible alternative.

Uneven load conditions on the wide decks can generate very high bending forces in the wide, squat pylons. In order to keep section sizes within the urban scale, and to allow for a subtle surface modelling, a unique system of integral composite-steel anchorage boxes faced in concrete was developed. The upper spires contain a very high-strength steel assemblage, carefully checked for the welding and fatigue problems associated with the strong but brittle high-grade steels, and these hold the anchorages in a very compact but accessible prefabricated box unit.

The very wide deck was subject to the weakening effect of 'shear lag'. Aeronautical engineers working on early monocoque wings and fuselages in the closing years of the First World War found that parts of very thin skin structures ceased to contribute strength if stringers were too widely spaced. This phenomenon appears in all lightweight box structures and must be carefully controlled to achieve an efficient design. The extent of deck section that was fully stressed was very difficult to determine, and therefore parametric sensitivity studies were undertaken to determine if the assumptions that had to be made were accurate, as they would be inordinately dangerous if incorrect. In the design phase different amounts of the overall section were activated under different load conditions, a process that offered a safe compromise to testing the entire deck.

The cables themselves experience unusually high bending stresses. A special evaluation was made and the parallel-strand cables were left ungrouted within polypropylene sheaths, for flexibility. The susceptibility of stays to wind excitation is a persistent problem with cable-stayed bridges, and the wide range of lengths and tensions makes it difficult to resolve. When a rod flaps, the anchorage points (even pin ends that cause small amounts of friction) restrain the movement and cause local bending, so that fatigue, the eventual loss of strength in metal caused by a large number of stress changes, sets in. In certain conditions rivulets of rainwater running down and around the cables can aggravate the problem, but no precise explanation of the phenomenon has yet been agreed on. The simple but unsightly solution is to tie the arrays together with tuning wire adjusted to tensions just sufficient to stop the onset of movement. Another method is to fit hydraulic or friction absorbers at the anchorage points, but traffic vibration compromises the life span and effectiveness of both these damping systems.

In the past a common site remedy for the problem has been to wrap a nylon rope temporarily along the length of the offending rod to disrupt the air flows. On the Charles River Bridge this idea becomes a spiral beading extruded along the surface of the plastic stay casings, while silicone-filled bladders within the casings, adjacent to the anchorages, replace unsightly 'sock' dampers.

The stay sheaths achieve one further degree of sophistication. The parallel strands are left ungrouted but are surrounded in rust-inhibiting grease that can be flushed through and replaced. Sunlight heats and liquefies this packing so that pressures build up and therefore, as part of the extrusion process, the high-density poly-ethylene is given a white outer face and a black inner surface, enabling the grease to stay cool and solid.

The landmark status of the Charles River Bridge, its confined urban siting and its construction constraints, which required the maintenance of a hectic transport system, has brought forth a whole range of technical innovations and refinements to an established system. A continual stress on ingenuity has also given the bridge a unity that the complexity of requirements threatened to preclude.

Sunniberg Bridge; Klosters, Switzerland; Christian Menn / Bänziger Köppel Brändli and Partner 1998

140m / 459 ft

The Sunniberg viaduct stands at the pinnacle of a distinct branch of European bridge-building. This idiosyncratic design is the product of the veteran Swiss bridge designer Christian Menn. Nuances of the individual designer can be traced alongside a number of specifically Swiss characteristics.

Menn's early schemes, in the 1950s and 1960s, appropriated and extended the work of French designer Robert Maillart. Close observation

of plaster test models confirmed their belief that reinforced concrete is a plastic rather than an elastic material. French designers had tended to treat concrete as a replacement 'wonder material', which behaved elastically in accordance with existing analytical techniques. To justify this assumption they detailed the concrete with relieving pins and joints. In contrast the Swiss approached concrete as a monolithic substance, gradually cracking and creeping but best cast into one unit

in which internal forces could adjust over time. The Sunniberg Bridge curves across a deep glaciated valley above the ski resort of Klosters, linking two tunnels in the cliff face. The curve on plan is exploited in several ways. Lateral stability comes from the deck arching horizontally between abutments. Support piers are restrained by the deck, rather than cantilevered, to resist sideways forces. This enables them to take a refined form to reduce visual intrusion. The harsh mountain conditions

impose a wide temperature swing on the structure. Expansion and contraction are accommodated by the deck springing into or out of the curved alignment. The piers taper towards their bases so that they can rock sideways without developing large bending forces.

To keep down the height of the structure above deck level the cable-stay configuration takes the form of a low, stiff harp. The parallel groups of cables have an infinite focus and therefore a natural repose.

There is little difference in structural efficiency between harp and fan patterns of cabling; parallel-strand and open-fan arrangements both avoid the bunching of anchorages at the head of the supports. The cables used at Sunniberg are parallel strands held within a protective sheath. Such cables are at the upper limit of efficiency in material use. Each wire can be hard drawn to its maximum strength. The brittleness that results from this process is mitigated by limiting the bending

forces and protecting the delicate strands within robust covers. However, cable corrosion remains a key concern in all suspension systems. In a parallel-strand cable the sheathing can be used as a duct, which is packed with rust-inhibiting material that can be flushed out regularly and replaced if necessary. In accordance with the Swiss emphasis on maintenance provisions, the member sizing on the Sunniberg Bridge allows for the removal of any single cable while the

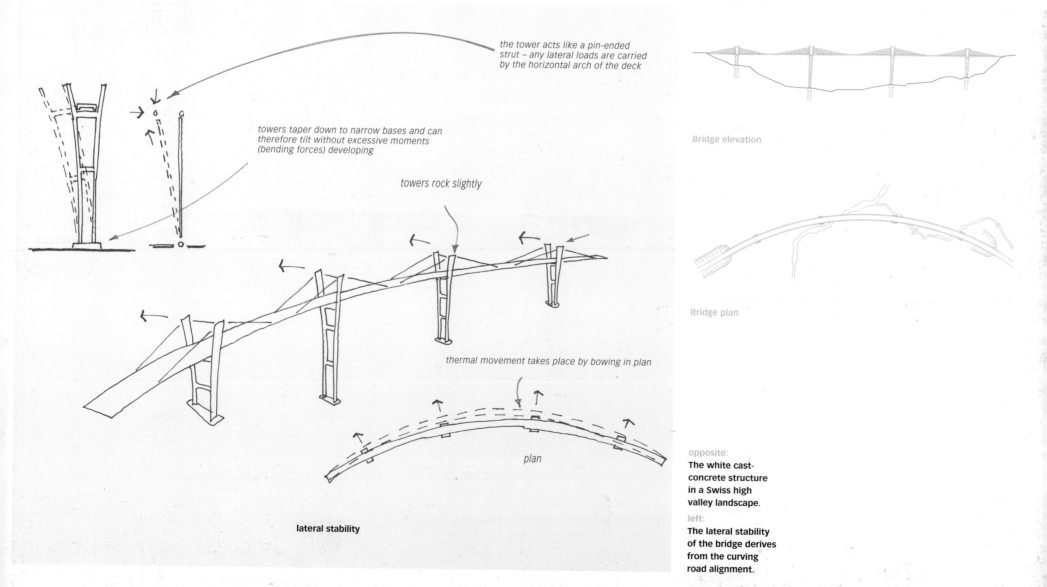

the tower acts like a pin-ended strut – any lateral loads are carried by the horizontal arch of the deck

towers taper down to narrow bases and can therefore tilt without excessive moments (bending forces) developing

towers rock slightly

thermal movement takes place by bowing in plan

plan

lateral stability

Bridge elevation

Bridge plan

bridge remains open to traffic. At the anchorages, tensions have to be removed evenly from each strand. This is done by spreading the bundle out among individual wedge sets. The conical block of the anchorages is bulky. The casing itself has to be lightly stressed if unsightly bends are to be avoided and the end grips increase the girth of the anchorage still further. At Sunniberg space has been created for the anchorages in the wide, sculptural pier heads. The pylons are modelled as a set of three planes, and these separate above the bridge deck to enclose a chamber behind the anchorage blocks. The point loads from the cable ends try to wrench the blocks from their sockets, but the splayed profiles provide sufficient concrete to contain the high bursting forces behind the spreader plates. Additional tie plates are fixed around the chamber walls to guarantee an even flow of forces. The anchorages can be readily inspected, and the drained space ensures that water cannot build up, a common problem with sealed anchorages.

Careful detailing minimizes the apparent depth of the deck structure. The cross-section is heavily modelled with a feather edge outside a pair of longitudinal stiffening beams set well back in the soffit shadows. The cables are anchored in the edge plates rather than onto the beam lines, giving the appearance of delicacy at the expense

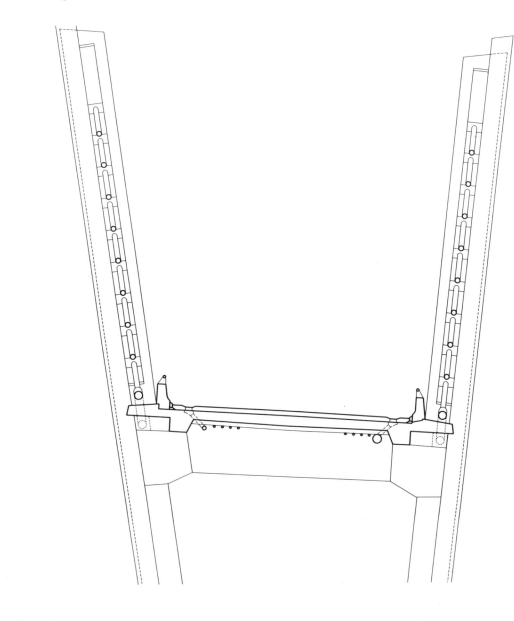

of increased reinforcement. Precast crash barriers form low parapets topped by simple and unobtrusive handrails. Road drainage is tucked away inside the beam lines, invisible but readily accessible.

The directness and the economy of means is a reflection of the local consciousness. The potential of structural details for visual communication remains unacknowledged: the main piers meet the ground without inflexion; the cables are allowed to slip between two planes of reinforced concrete into a recess hiding their anchorages.

The design could be seen as subject to a tension between incompatible ideals. Technical measures have been taken to reduce the impact of the structure on the landscape, an ideal traceable to the attenuated, trabeated viaducts built as part of the German autobahn system, where no effort is spared to reduce the size and complexity of each member to a sparse composition of simple forms. At the same time there is an acknowledgement of presence, a sensibility that can be traced back through Menn's work.

The Sunniberg Bridge handles the relationship of cable groups to curved deck in a more ambivalent way, revealing the maturity of this designer's work. The upper pier sections curve outwards to keep the unconfined cables clear of the road line. The open arrays of cable are visually unobtrusive but the sculptural pier heads draw the eye.

opposite, left:
The pier heads splay outwards to tilt the cable fans away from the road alignment.

opposite, right:
The apparent depth of the roadway is reduced by setting the longitudinal beams inboard behind a precast edge profile and placing the roadway services under the central soffit.

left:
Menn's Ganter Bridge, Eisten, Switzerland, 1980. An expressive structure is derived from another method of cable-staying a curved bridge. The hangers are enclosed in curving panels of concrete.

Solferino Bridge; Paris, France; Marc Mimram 1999
102m / 335ft

The close connection of Paris with the River Seine generates a unique cityscape. More than 30 bridges take their place in this milieu and the introduction of a new crossing into such a highly developed context provides an opportunity to examine its nature.

The Solferino Bridge replaces a temporary steel structure erected on the site of a dilapidated landmark bridge between the Pont Royal and Pont de la Concorde. Conceived in the tradition of the French grands projets – masterpieces celebrating social, national and cultural achievement – this pedestrian bridge connects the Jardin des Tuileries and the Musée d'Orsay. Following an international competition the project was awarded to one of France's most prominent engineer-architects, Marc Mimram, whose stated aim was to create 'a contemporary design fitted to the historical and symbolic context'. The structure takes the form of a low, arched bridge with a promenade deck. Each end has a high and a low access point. The complexities of the site are beautifully resolved into a perfectly symmetrical bridge without staircases. The upper deck rises gently from each side, passing around ramps that run back down the arches to the river edge. The principal local route, from the low-level foot tunnel beneath the regional express station on the Right Bank across to the Left Bank, fits smoothly into this arrangement. The intersection of paths at the

centre of the bridge makes it a stepping off point from which to encounter the city.

The ramps pass between two arch rings composed of cast segments. In the competition scheme these voussoirs were aluminium-cube frames friction-welded together, suggesting an avant-garde interest in innovative materials and techniques. Weak welds were considered possible because the arch form keeps the joints in compression at low stress levels. However, high costs and concern about the material's flexibility (aluminium is only one third as stiff as iron) led to the adoption of steel throughout.

Mimram's declared fascination with organic structure is indulged in the over-jointed arch system, with its repeated references to vertebra. The broad deck is supported on spandrels of filigree framing, and in turn braces the arch rings below and counteracts their tendency to shed load by buckling sideways. Such very low arches have a propensity to buckle vertically, to 'snap through' and hang between the bearing points. Very rigid foundations help to resist this inclination.

There are numerous references within the design to the nearby Pont des Arts, a supremely elegant exposition of urban, lightweight wrought iron, and to the other bridges along the Seine, with their low, segmental arches and pierced spandrels. The characteristically

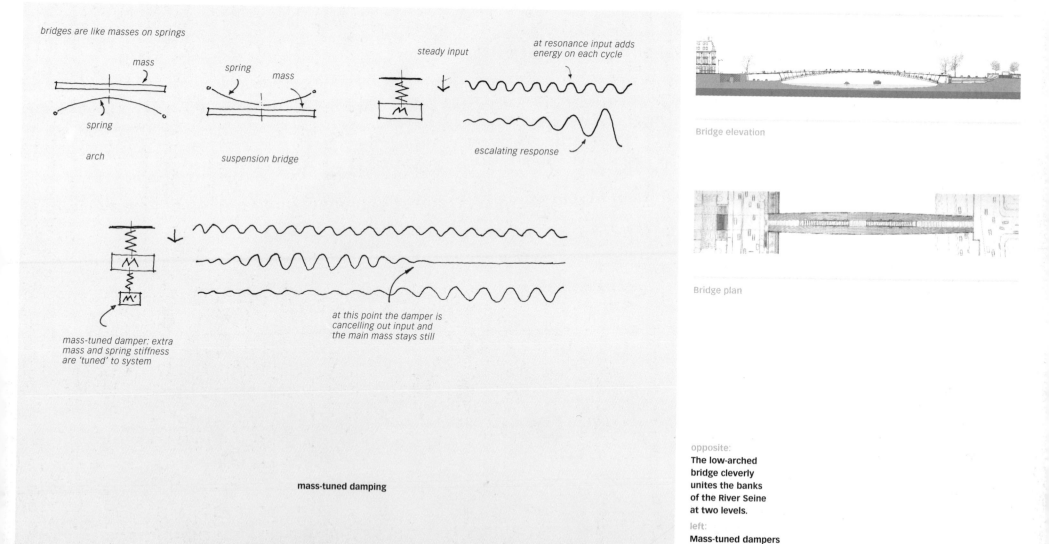

bridges are like masses on springs

mass

spring

mass

spring

mass

arch

suspension bridge

steady input

at resonance input adds energy on each cycle

escalating response

mass-tuned damper: extra mass and spring stiffness are 'tuned' to system

at this point the damper is cancelling out input and the main mass stays still

mass-tuned damping

Bridge elevation

Bridge plan

opposite:
The low-arched bridge cleverly unites the banks of the River Seine at two levels.

left:
Mass-tuned dampers vibrate to cancel out pedestrian-induced excitations.

Parisian use of trellised metal construction, found in the glass roofs, arcades and canopies nearby on the Right Bank, and in the Eiffel Tower, is invoked in the braced rings and strut supports. (Appropriately, the company founded by Alexandre Eiffel built this structure.) The bridge's complex arrangement of metalwork continually reflects light onto the water below.

Unfortunately, the Solferino Bridge was closed on the day it opened because the deck shook too much. All bridges move but the critical factor for a footbridge is whether the movement is perceptible and, if so, whether it is merely discomfiting to the user or, more dangerously, enough to make pedestrians fall into a rhythmic pattern of movement. If accelerations of the deck are perceptible, sensations like those experienced in a very fast elevator are induced (vertical movement) or there is unsteadiness when moving along (horizontal movement).

The human stride has a frequency of about 1.5–2Hz, one or two footfalls per second. Bridges of medium size built to be as light as possible, typically spanning highways or rivers such as the Seine or Thames, have natural frequencies close to this level. The pedestrian therefore taps the bridge at its frequency of vibration. The energy added on each cycle accumulates and, if it is not dissipated, movements in the bridge build up.

Unusual structures must be tested to see if the assumptions made in their design are accurate. Transducers placed on the deck describe and record the bridge's movement through space over time, but

despite this the testing process can become a classic 'black box' problem: it is necessary to identify the structure's characteristics but calculating a suitable input of force is difficult in big bridges. Artificially induced forces must be large and are therefore expensive to apply. Natural sources are a viable alternative, however, as the wind operates on a broad but characteristic spectrum.

The true natural frequencies of a bridge can therefore be identified and the mode shapes, the curves and waves into which it twists, can be plotted. Each shape, whether vertical, lateral or torsional, has its own harmonics, multiples of the simple waveform, like a stringed instrument. Once these are known the damping that needs to be mobilized within the structure for each of these patterns can be determined.

On the Solferino Bridge the arch structure is ideally suited to carry extra load, and the vibration modes of its symmetrical configuration proved measurable. The visual complexity of the design is robust enough to support the addition of mass-tuned dampers, and the lightweight construction means that these moving weights are not unduly large. After 25 tonnes (24.6 tons) of mass-tuned dampers had been fitted the bridge reopened.

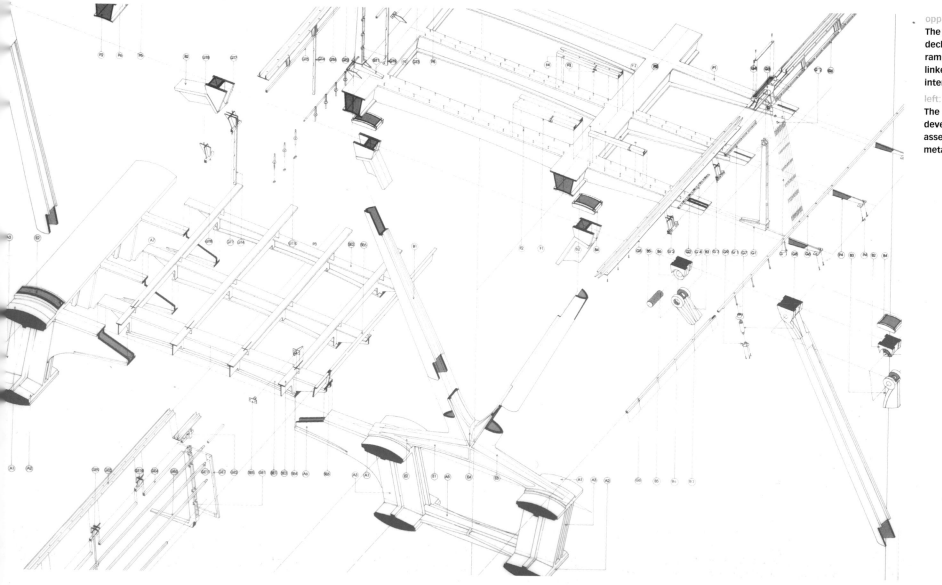

opposite:

The arch and upper deck incorporate ramps and walkways linked by a central intersection.

left:

The structure was developed as an assemblage of cast-metal elements.

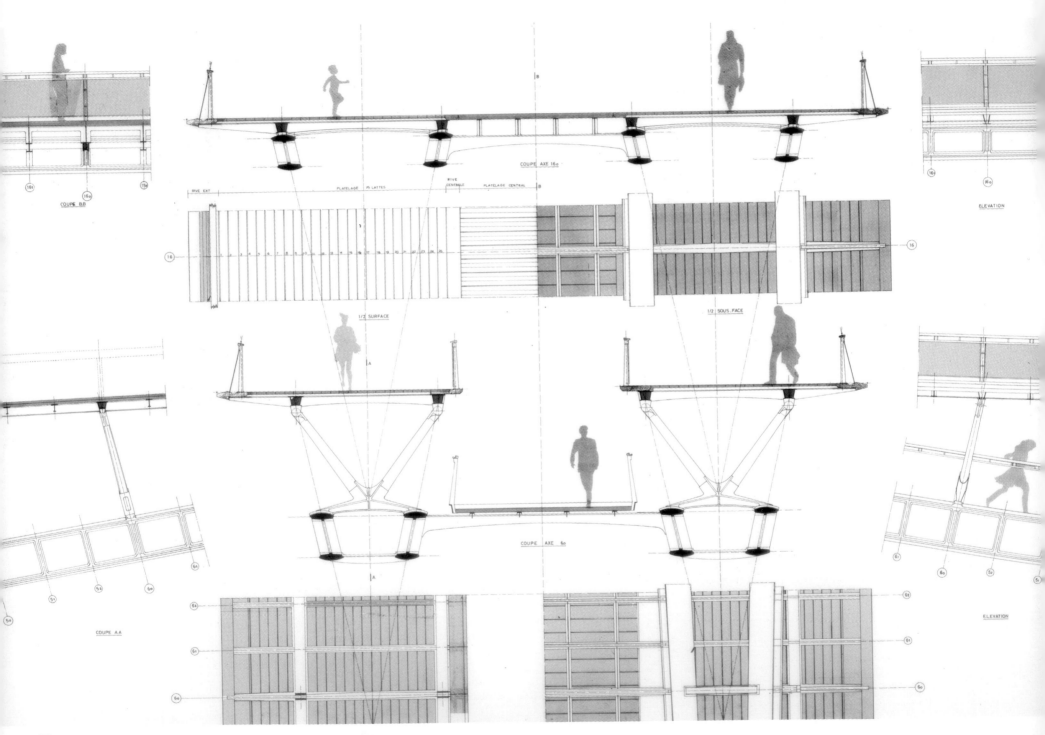

COUPE AXE 16o

RIVE EXT   PLATELAGE 25 LATTES   RIVE CENTRALE   PLATELAGE CENTRAL   B

COUPE B.B

ELEVATION B.B

1/2 SURFACE   1/2 SOUS .FACE

COUPE AXE 6o

COUPE A.A   ELEVATION

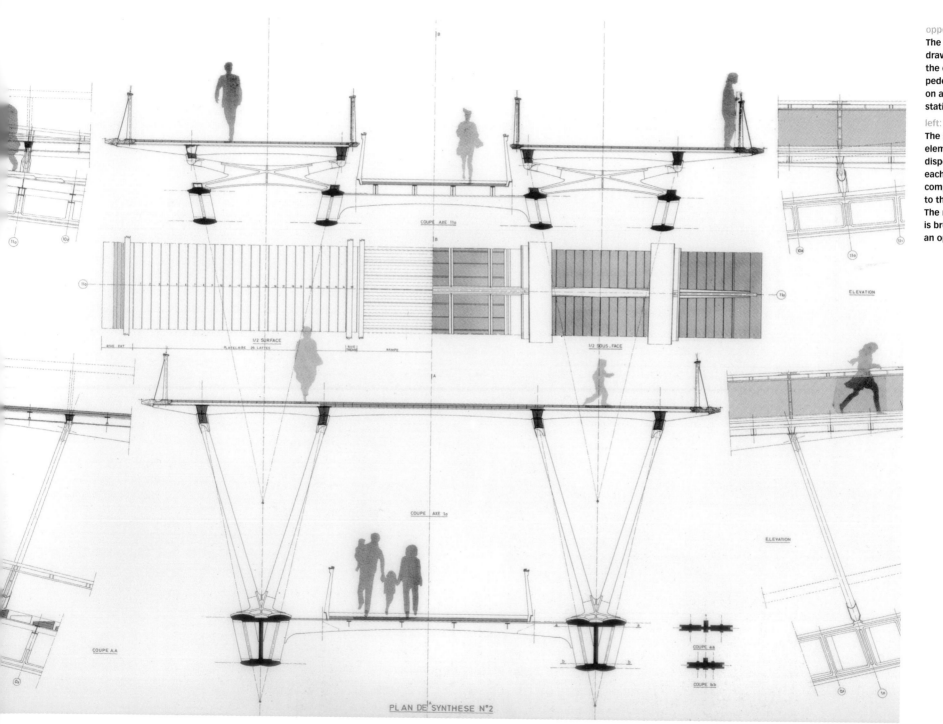

opposite:
The architects'
drawings investigate
the experience of
pedestrians moving
on and within the
static form.
left:
The structural
elements are widely
dispersed so that
each individual
component relates
to the human scale.
The main arch ring
is broken down into
an open box-work.

COUPE AXE 11o

ELEVATION

1/2 SURFACE

1/2 SOUS .FACE

RIVE EXT

PLATELAGE 25 LATTES

RIVE TREME

RAMPE

COUPE AXE 1o

ELEVATION

COUPE A.A

COUPE aa

COUPE bb

PLAN DE SYNTHESE N°2

Sunshine Skyway Bridge; Tampa, Florida, USA; Figg and Muller 1986
365m / 1198ft

The acceptance of innovations in bridge design as standard practice follows a pattern: new materials are assessed in terms of those they replace; new construction methods reconfigure available means; new structural concepts come about as a recombination of accepted models; new analytical techniques retain and refine systems that have proved too blunt. The Sunshine Skyway Bridge incorporates several developments and refinements made by the French engineer and inventor Jean Muller, reflecting the important and varied contributions he has made to the industry as a whole. Now one of the most well-known and popular modern bridges in North America, the new Sunshine Skyway arose from a disaster. The site had been occupied since 1954 by a pair of through-truss bridges, which had the classic southern American silhouette. On 9 May 1980, in a severe rain squall, a Tampa Bay pilot on board an empty freighter missed his turn on a channel marker and careered into a centre-span support. One main span came down, taking 36 motorists with it. The incident led to a completely new project for a single bridge. The second Sunshine Skyway is the largest cable-stayed bridge in North America. Its aesthetic and engineering are based, respectively, around the reductionism and innovations of the French engineer Eugène Freyssinet. Muller was Freyssinet's protégé and attributes to him the words that became his guiding ethos: 'a resolute desire to simplify form'. This emphasis on the minimum of means underpins much engineering output; the idea that valuable resources should not be

wasted is transformed into the precept that good design flows directly from economy.

The bridge rises in a long convex curve across the waterway, emphasizing the transit of space. The deck is developed as a continuous ribbon of precast-concrete segments. Freyssinet pioneered the process of prefabricating large bridge elements in concrete to avoid expensive shuttering. Muller took a leap forward with 'match casting': by casting each piece against its neighbour the joint between the two is perfect; it is exactly the fine butt joint needed for resin glueing to work.

The approach ramps are simple prisms of reinforced concrete set on elliptical piers and post-tensioned together. As each segment was added during construction they were pulled together with tendons threaded through cast-in ducts. The manipulation of forces made possible by this process of post-tensioning, or stressing after casting, allowed Muller to use another of his construction innovations, gantry launching.

This method involves the repeated jetting forward of segments onto the end of an ever-growing cantilever, stiffened by a temporary truss, until it eventually reaches the next pier. The structure goes through several stress compositions during its construction: moments, shears and axial forces ebb and flow as work proceeds.

Muller brought the system to a new level of perfection on the Linn Cove Viaduct in North Carolina, which was being constructed when the Sunshine Skyway Bridge was in design. In this project he installed

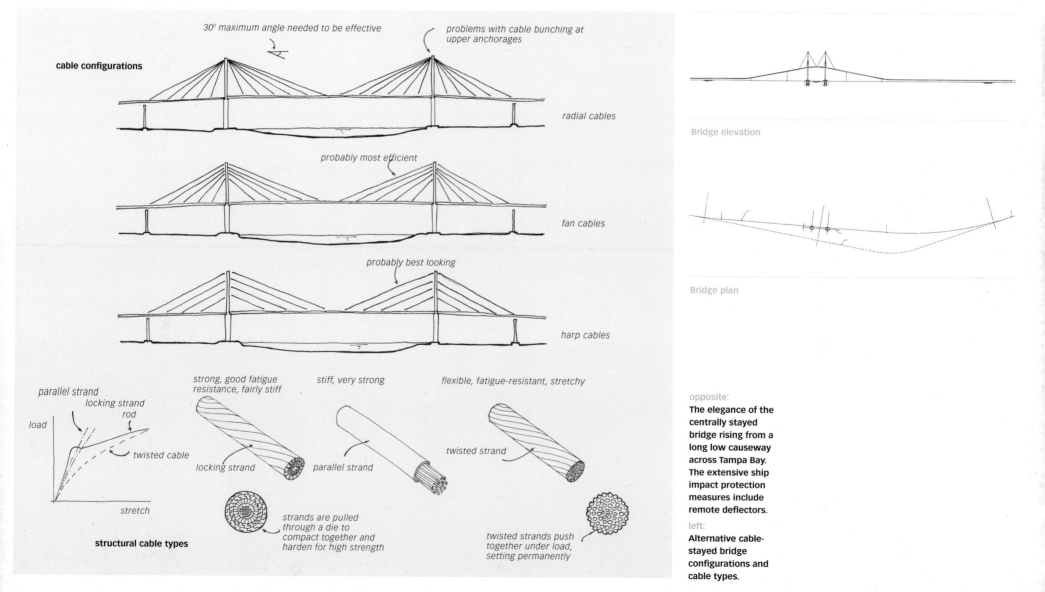

cable configurations

30° maximum angle needed to be effective

problems with cable bunching at upper anchorages

radial cables

probably most efficient

fan cables

probably best looking

harp cables

parallel strand

locking strand rod

load

twisted cable

stretch

structural cable types

strong, good fatigue resistance, fairly stiff

locking strand

strands are pulled through a die to compact together and harden for high strength

stiff, very strong

parallel strand

flexible, fatigue-resistant, stretchy

twisted strand

twisted strands push together under load, setting permanently

Bridge elevation

Bridge plan

opposite:

The elegance of the centrally stayed bridge rising from a long low causeway across Tampa Bay. The extensive ship impact protection measures include remote deflectors.

left:

Alternative cable-stayed bridge configurations and cable types.

a complete viaduct from deck level. As the bridge crept out above the ground its supporting piers were installed downwards from the deck. There was no access at ground level at all. In this way he avoided damaging the hillside's sensitive slopes and causing the construction scars that can mar a landscape for decades.

Muller's confident innovation developed from a background of practical experience in a particular construction method. The relatively new technique of prestressing, particularly in precast systems, is not yet hemmed in by received knowledge. Freyssinet and Muller have established a standard of practice in this area, a canon of works, although their expertise is in the system itself, not in building types. Freyssinet has completed airship hangers, the most marginal of forms, while Muller has explored the potential for concrete oil tanker hulls and built atomic reactor vessels for the French nuclear programme. Such a breadth of application leaves the engineering mind open and responsive.

As the Skyway stretches across the middle span it is picked up on a central line of cable stays, set in fan arrays from two substantial pylons. The spine of cables allows uninterrupted views across the bay and reduces wind loads. A nearby weather station measured the highest wind speed ever recorded, 306kph (190mph) during Hurricane Camillo. The bridge elements are therefore reinforced to sustain a steady wind of 386kph (240mph) and a gust of 467kph (290mph).

The stays carry vertical loads, while sideways forces and torsions from unevenly loaded carriageways are taken by the stiff, concrete deck girder. The balancing of cable stiffness and deck stiffness in a cable-stayed structure gives the form much of its character. On the main span the box girder is stiffened with internal struts, complex

assemblies built up by sequential precasting. Using a continuous, concrete box-girder deck gives a very stiff deck element. As a result the cables do not have to be as widely dispersed as on steel bridges, which are more flexible. Muller has argued shrewdly that the acceptable level of stress can also be revised upwards. The tensile strength of cable stays is set by precedent. Above a stiff deck, cables are not subject to the wide stress cycles sustained in older deck forms, and they should therefore be able to sustain higher long-term loads. Following the ship strike a sophisticated statistical analysis of potential impacts was made, now established as a general method for assessing the problem. The new bridge is protected by massive outboard fairings. The groups of piles forming the foundations are locked together with prestressed-concrete capping rings, which mobilize the clusters to act as integral units resisting collisions.

The refreshing directness with which the handful of components are arranged, the overall simplicity of the elements and the natural proportions of the structure are perhaps compromised by the casual treatment of the intersections. The way in which verticals should meet, pass and support the horizontal, the treatment of base intersections and apexes, is a problem that has always challenged designers. In the Sunshine Skyway the cables meet the pylons in an elegant recess, but otherwise the need to acknowledge salient aspects of the form has been overlooked, perhaps because an architect was not employed to address the overall design. In the fluency of the construction the formal way in which elements are put together needs a voice.

The upper anchorages of the main cable array are set neatly into central recesses on the profiled pylons.

Modern and traditional bridge configurations adjacent to one another invite comparison.

Main-Danube Channel Footbridge; Essing, Germany; Richard Dietrich/Heinz Brüninghoff 1986

73.28m/240ft

Underlying the construction of this large-scale timber bridge – an exercise in sustainable building – is a close engagement with the environment and landscape. The site is in a broad, flat river valley, the Altmühltal (Old Mill Valley), between steep wooded slopes, where the ancient castle and picturesque village of Essing occupy a rock formation around which the river once turned. A path has been opened here across to a nature reserve. The walkway, wide enough for light vehicles and farm machinery, crosses several features: an old

lane, the wide navigation of the canal, a second lane, an oxbow lake left by the river, and the new main road. The continuous profile of the deck, set up on narrow stick supports, relates to the extraordinary geometric arrangement of this landscape, bisected by the canal. The brief called for an old timber bridge to be taken into account in the design of the new structure.

The designer, Richard Dietrich, places his work within a continuous development of bridge design, a series of master works building into

a new sensibility. He prides himself on living between the urban and technical sophistication of an office in Munich and an isolated farmhouse in Cheimgau. This design appropriates nature and uses it as a metaphor, taking up an aspect of German Romanticism.

Practicality becomes secondary to the experience of the bridge. The structural form is a continuous, curving profile emphasized in its detailing. The undulating profile of the walkway harmonizes with the strolling movement of recreational walkers. The ribbon structure is a

minimal intrusion in the landscape, and the cheap, galvanized-mesh balustrade panels are virtually invisible at a distance, further attenuating the bridge's line. This simple design statement is on the correct scale to address its surroundings. The organic form and the suppression of the main joints hint that the structure might have grown into place.

To realize these effects glued laminated beams are draped across a series of fans, a motif lifted from far older trestle forms, and restrained by anchor blocks at each end. Continuity is stressed by the heavy, built-up deck line underpinned by the stick and cable supports. The tensions that develop in the sagging spans reduce the structural depths; 65cm (26in) depth of solid timber is used, as opposed to the alternative 4m (13ft) height of open truss. The main supports are paired sets of braced struts. Locked into the beams above, they act as stable frames longitudinally and reduce the rotations that would enlarge deflections under uneven load conditions. The inner props are subtly recentred to improve rotational stiffness and enhance the system's resistance to vibration. The first engineer employed on the development said that it couldn't be carried out, having visualized the proposition in the context of inappropriate precedents. Working from more fundamental principles, Heinz Brüninghoff was able to stay with the designer's concept.

Timber engineering is predicated on the handling of three factors: wood's variability, its movement behaviour and its jointing characteristics. The first factor is dealt with here by the standard

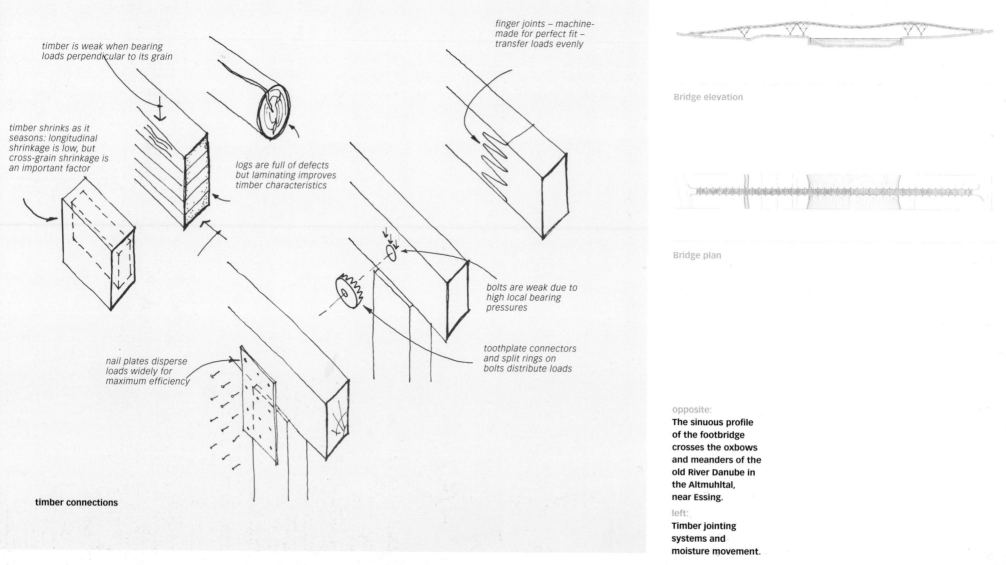

timber is weak when bearing loads perpendicular to its grain

timber shrinks as it seasons: longitudinal shrinkage is low, but cross-grain shrinkage is an important factor

logs are full of defects but laminating improves timber characteristics

finger joints – machine-made for perfect fit – transfer loads evenly

bolts are weak due to high local bearing pressures

toothplate connectors and split rings on bolts distribute loads

nail plates disperse loads widely for maximum efficiency

timber connections

Bridge elevation

Bridge plan

strategy of accepting a high level of redundancy within individual elements, employing multiple bolt connections and laminated components, and, within the overall arrangement, using parallel beams and supporting struts ranked across its width. The main longitudinal bands are tightly laminated baulks of timber. Growth limits the size of elements to short lengths and small sections but also introduces flaws, imperfections and inconsistencies. To reduce the influence of defects in any single piece of timber the bridge uses built-up members, here prefabricated to the enormous length of 45m

(148ft) and transported by road. The main supports are viewed as secondary constructions and take the form of single squared elements joined by direct and expressed bolted joints. The variations in the single wood sections are accommodated by stress grading, by rating the wood within a band of performance. The sections inevitably become heavier to cater for this approximation. The handrails are resolved as an additive element, carefully slanted and detailed to shed water. The lateral resistance of the bridge is provided by cross-bracing comprising sticks planted on the soffit of the main beams taking loads

to tie rods laced across and between the supports. Timber expands and contracts as the moisture content of the air changes, and over time it usually dries and shrinks. Cross-grain shrinkage is four or five time as great as longitudinal movement and quickly loosens joints. Over long periods stressed timber creeps and relaxes towards the load being applied. The draped form of the bridge is the ideal type to cope with these changes. It sags slightly as time goes by, and therefore no movement joints are required. The structure's jointing centres around strategies for taking the forces evenly distributed in a timber

section out of the timber fibres and redirecting them appropriately. The design employs a mixture of jointing techniques, applied appropriately and conservatively; there is little innovation in an area where much development might be expected. The main site joints are mechanical. Loads pass from distributed groups of nails into steel plates and across metal-to-metal connections concentrated in pins or bolts. Prefabricated joints are made with glued finger joints that transfer the load from grain to grain. Traditional scarf joints are also employed. The main supporting struts are finished with cast-steel

nodes, which carry the prop ends, protect the cut grain and channel forces through pins onto the centre lines of each part. Design concerns mainly revolved around the bridge's dynamic behaviour. Spruce, the base material, is light and strong. It is fast-growing, and therefore cheap, but also highly elastic. The structure was necessarily going to be very springy. Analysis showed that the very long continuous elements have natural frequencies remote from the human footfall. Deliberate swinging was not foreseen to be a problem as the perceptible modes of movement have a large mass participation; for

people to set the bridge vibrating they would have to move considerable portions of its weight. Damping, or the ability to shed energy, is high in timber structures. Wood is elastic but dissipates energy within itself when it is flexed. The lapped joints fret, rub and grind together when the bridge sways. This friction generates heat and prevents vibrations. The construction sequence was traditional: prefabricated subsections were set on prepared piers. Jointed by familiar means, the constructed bridge becomes a conjunction of idiosyncratic, original thought and proven conventional systems.

opposite, left:
The main support's base detail. Steel castings are shaped to integrate the function of pin ends, supporting blocks and providing end grain protection to the principle timber members.

opposite, right:
Handrails and balustrade posts are detailed to shed water, preventing movement caused by rot and undue moisture.

left:
The unrelenting profile of the main laminated beam is broken down by the built-up balustrade assembly and horizontal soffit bracing.

143

Roosevelt Lake Bridge; Phoenix, Arizona, USA; HNTB 1990
329m/1079ft

Roosevelt Lake, above the city of Phoenix, Arizona, was formed in 1911 by the construction of a masonry gravity dam in a narrow gorge on the upper creeks of the River Salt. As the reservoir gradually choked with sediment it became necessary to raise the old dam, a deep concave wall of massive limestone blocks, to prevent flooding. This also involved altering a narrow road that runs along the rim, connecting the mining towns of Globe and Payson. In addition, the area's development as a water sports centre was hampered by the road alignments. The

haulage road from Phoenix and Mesa used during the dam's construction, known as the Apache Trail, had been appropriated for recreational use. It was, however, built for horse-drawn transport and included hairpin bends that could not be negotiated by boat trailers. The option of realigning and broadening this road so that it ran across the dam was examined but rejected in favour of bridging the gorge above the barrier, where the valley opened out, and taking both roads across the bridge. The long span of two lanes makes the bridge

unusually slender. The geological conditions that had determined the dam's location suited a big bridge solution. A viaduct was considered but the piers would have been lost in the deep water and alluvium beneath the lake. The extensive survey work carried out when the dam was commissioned, and photographs of the excavations during construction, provided precise information on the ground conditions on either side of the valley. The rocks dip steeply but the Redwall limestone is strong and could readily sustain foundation loads from

tall towers or arch abutments. Several schemes were developed in parallel, and the final choice was made on the basis of bids received to do the work. Economic comparisons are difficult to make because different programme times and maintenance regimes complicate the assessment. The steel arch chosen had the lowest base cost, proving competitive with cable-stayed concrete schemes because the lateral forces generated by the arch form could be sustained by the hard rock walls on each side. Furthermore, the roadway did not need reinforcing in order to act as a tie and was reduced to a light steel and concrete composite deck that could be lifted into place off lake barges.

The designers claimed that the choice was entirely pragmatic, however the scale and presence of the arch resonate perfectly with the spectacular site, providing a landmark for the waterpark. The heyday of this well-proven form coincided with a heroic period in American bridge engineering in the first half of the twentieth century. The bridge therefore takes its place in the context of American Midwest conservatism.

The structure is very simple, and the detailing is direct and unselfconscious, including junctions that are elemental and roomy. Both structure and detailing are imprinted with the contractor's influence. The design analysis was as simple as the size of project allowed – a good way of avoiding miscalculations and oversights. Since the layout is straight and symmetrical the elements were sized using a plane-frame analysis (only two dimensions). Large deflection theory was used to assess the bridge with this type of analysis. When

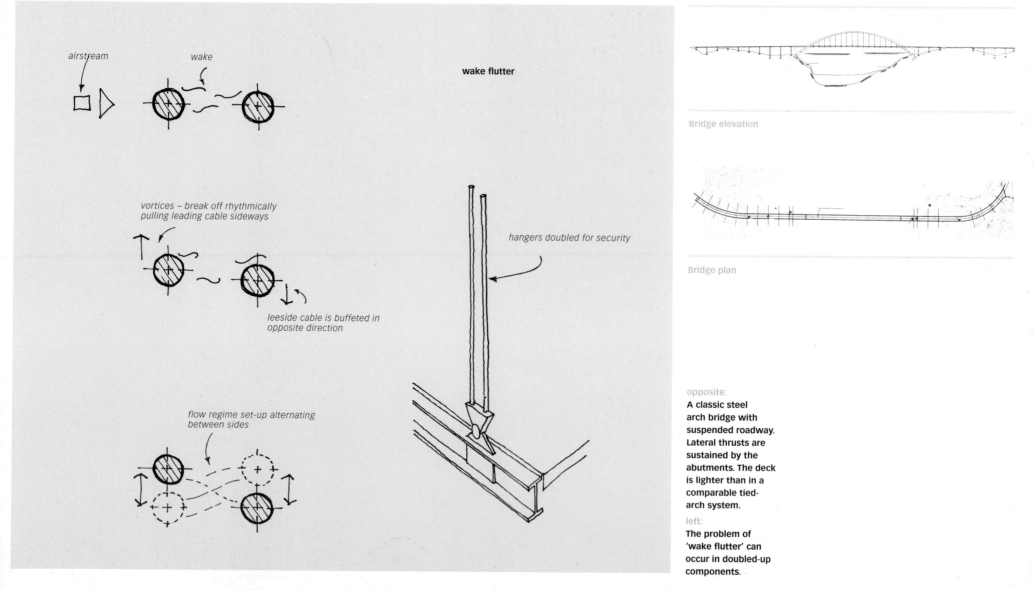

airstream   wake

wake flutter

vortices – break off rhythmically
pulling leading cable sideways

hangers doubled for security

leeside cable is buffeted in
opposite direction

flow regime set-up alternating
between sides

Bridge elevation

Bridge plan

opposite:
**A classic steel
arch bridge with
suspended roadway.
Lateral thrusts are
sustained by the
abutments. The deck
is lighter than in a
comparable tied-
arch system.**

left:
**The problem of
'wake flutter' can
occur in doubled-up
components.**

the structure distorts out of shape the alignments of forces shift within each element. This generates substantial additional stresses, and the computer must be carefully programmed to take these reconfigurations into account. A seismic analysis and wind-loading check were also carried out, on a three-dimensional computer model. Results from the simpler, two-dimensional model could then be used to validate the three-dimensional analysis. A wind tunnel was used to test the design for aerodynamic effects. The bluff edges of the box elements make for

easy fabrication but shed vortices in an airstream. However, the open crosspieces break up rhythmic flows, and no countermeasures were required on the main structure. The hangers were grouped in sets of four bars, and these needed dampers to preclude wake flutter, in which eddies breaking off one bar force another in its lee to resonate.

The paired arches need crosspieces to prevent the ribs buckling sideways and to resist the high winds funnelled down the sides of the lake. K-braces, triangulated openwork frames, proved far more efficient

than using the less visually intrusive Virendeel bracing, in which boxed cross-beams form a ladder-like pattern of rectangular panels.

Elementary prestressed-concrete beams on reinforced-concrete hammerhead piers are used for the approach causeways. The ideal parabolic profile of the upper arch kinks at the deck line to take the additional load from the side spans. The upper works are steel box girders stiffened by stringers and diaphragms and sized for walk-through inspection. The bases, often submerged, are prestressed-

concrete sections reinforced with bars coated with epoxy resin to increase longevity.

The arch was erected by cantilevering out the ribs to meet on the centre line. The concrete base sections on their spread footings were cast in place with a toe-piece behind. These elements were first of all loaded with kentledge (balancing ballast) to prevent the ribs from overturning and then became the base for tall temporary towers. The ribs were extended out, section by section, and held back by stays. To keep the pattern of forces in the temporary arrangement one stay was used on each side and progressively relocated as work proceeded. The rib bases were arranged as pins in their temporary condition so that the upper ends could be readily pulled into alignment to meet at the centre. When the ribs were finally closing together the baseplates were locked in place and the lower rib sections prestressed.

The steel arch is painted powder blue, a standard shade of zinc-rich primer, but the structure is otherwise unadorned. The artlessness of the bridge sits well with the neighbouring dam. Both artifacts, separated by three generations of engineering development, share a certain kind of nobility emanating from purely utilitarian forms placed in remote settings.

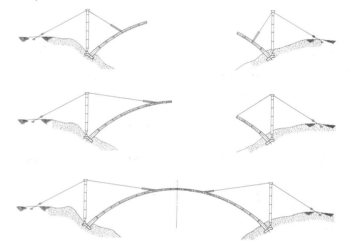

opposite:
The road alignment transforms the old packhorse trail it replaced and allows those driving campers and cars with boat trailers to exploit the recreational potential of the lake.

left:
The arch rings are tied by K-braces: inelegant but efficient.

above:
Final element sizes took account of the construction sequence. Temporary works were reduced to a minimal, easily manageable stay system.

Miho Museum Footbridge; Shiga Prefecture, Japan; I.M. Pei Architect/Leslie E. Robertson Associates 1996
120m/394 ft

The Miho Museum complex, buried in the Shigaraki Hills near Kyoto, is one of the principal designs of the veteran architect I.M. Pei. Embedded in the landscape, and in tradition, the galleries display an eclectic range of ancient art. The building is set deep within the summit of a hill, revealed only by a scattering of pavilions and other surface markers.

The site is isolated by the surrounding nature reserve and therefore required an approach that extended from a remote access point and passed through the hillside. This necessity was exploited to create a structure that is part of the visitor's experience. The extended approach is designed as a sequence of encounters with the landscape before the visitor is enfolded in the underground museum. The ancient Chinese literary device of an arduous journey ending in scholarly enlightenment at the pilgrim's temple precinct is reified in the sequencing of public access. Pei himself drew an analogy with the tale of the 'Peach Blossom Spring', in which a lonely fisherman walking in the mountains comes upon a grove of peach trees leading to a cave. Beyond is a land of enlightenment and plenty. After spending an afternoon there he returns to the world, but he can never find the grove again.

In Pei's design a winding path or golf-cart ride leads up from the car park to a small tunnel entrance, the 'cave'. The curving passage

beyond is lined with stainless steel, which gradually reveals the approaching light. The tunnel breaks out onto a footbridge across a deep, precipitous ravine. On the far side the museum site opens to one side. The Japanese design device of *shakkei*, a framed narrative of short and long views, is used to lead the eye away from the bridge and into the receding complex of buildings. The bridge itself is a hybrid of three mechanisms arranged to manipulate the pedestrian's perceptions.

The structure is not so much seen as experienced. Its form has been played down to make it as unobtrusive as possible, with the aim of immersing the visitor in the natural beauty of the site.

The single span has no piers, only abutments integrated into the walkway springing points, which lessens the impact on the environment. On the tunnel side an array of suspension cables is anchored to the tunnel lining so that the bridge is symbolically rooted in the mountain.

In Pei's eyes the mutual support of tunnel and bridge unites two disparate elements into a narrative sequence. The cables spring forward and are taken around a splayed parabolic arch, as in a giant violin bridge, before spreading down each side to the deck in a parting veil. Halfway across the span the stays disappear below the line of the deck balustrade. The remaining section is supported on a simple truss that is kept out of view below the walkway. A tetrahedral tubular girder

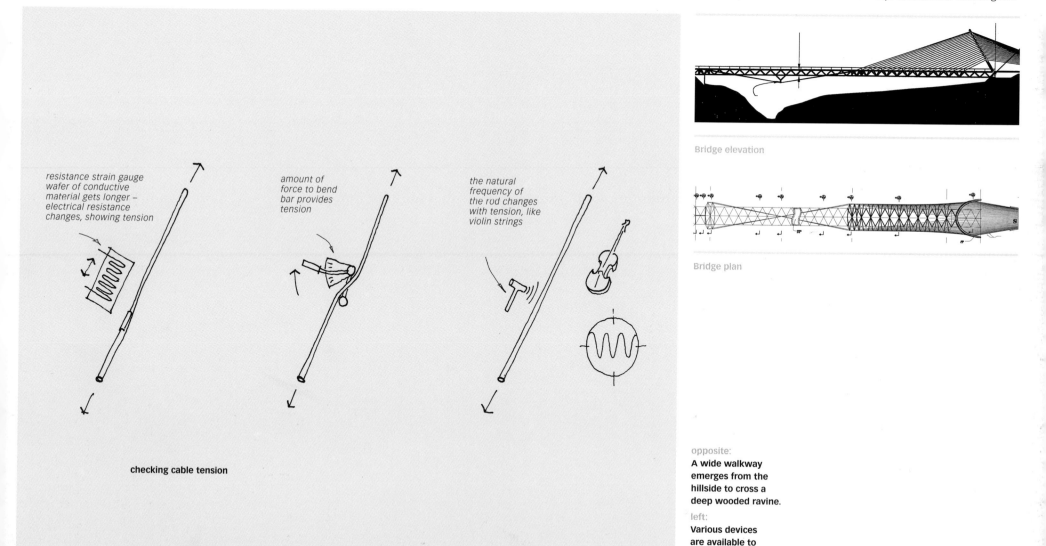

Bridge elevation

Bridge plan

resistance strain gauge wafer of conductive material gets longer – electrical resistance changes, showing tension

amount of force to bend bar provides tension

the natural frequency of the rod changes with tension, like violin strings

checking cable tension

opposite:
A wide walkway emerges from the hillside to cross a deep wooded ravine.

left:
Various devices are available to determine the tension in stressed cable systems.

149

stiffens the various parts. The balustrades are perforated-steel infills, providing a high level of physical and psychological protection to users. The geometry of the strainer arch and cable fans allows the bridge considerable freedom to change shape under load, and deflections are readily perceptible. This effect is intended to be part of the experience, much as rope bridges move when travellers edge across them. Dangerous vibration is precluded by the high level of internal damping and the complex non-linear behaviour of the distorting cables. The structure shakes in an unusual way, however, bending in response to loading and absorbing the energy produced by wind and walkers.

Maintaining this asymmetrical system's stability and repose required ingenuity. The cables are sized to suit the distribution of 'at rest' loads induced by the layout. Individually prestressed, they gradually grow in diameter towards the apex of the inclined ring. This pre-tensioning introduces a slight camber, an upward bow reminiscent of older footbridges.

Stressing a lightweight footbridge can be both expensive and complicated. Individual cables are adjusted by rotating a rigging screw or turn-buckle to alter the length. The leverage of a helical thread allows this to be carried out by hand on all but the thickest bars. In a

complex system the alteration of any one bar affects all the others so that each change 'cascades' through the other elements. Imperfections, friction in the joints and variations in cable diameters and stretch make these effects difficult to predict precisely. The engineer usually has to develop a tuning sequence by experiment to harmonize the loads in stages, but a difficulty arises in measuring the load in each bar. The expensive solution is to fix strain gauges to every bar. These sensitive

wafers measure minute changes in electrical resistance as the rod or cable strains, but they are delicate and difficult to calibrate. A less accurate method involves bending each bar locally by a very slight amount and then working out the tension from its deflection. At Miho the method used was particularly appropriate for a cable-stayed bridge. Each bar was twanged and, as with a guitar, the frequency at which it hummed betrayed its tension.

The environmental impact of the installation at Miho has been rigorously minimized. Bridges in the deep-wooded ravines of the Pacific Rim concentrate rain run-off around their piers or deck scuppers. The deck of this bridge comprises steel gratings filled with porous ceramic beads. These allow rainfall to permeate the deck and continue down into the forest canopy below with the same even spread as rainfall. The structure therefore becomes ecologically 'invisible'.

opposite:
The bridge is firmly rooted into the hillside. The main cable-stays cross a strainer arch and are anchored back to the tunnel lining.

left:
All the structural elements are arranged to reinforce the progression of the route.

Ushibuka Bridge; Kumamoto, Japan; Renzo Piano Building Workshop/Ove Arup and Partners 1995
150m / 492 ft

The career of the Italian architect and engineer Renzo Piano spans the development and aftermath of the high-tech phase of modern architecture. A prolific builder and co-author of the Pompidou Centre in Paris, his studies of structural form have been seminal. His earlier projects relentlessly pursued the generating principles of structure, environmental servicing and a preoccupation with detail and tectonics.

The critic Renato Morganti describes Piano as 'disquietingly inquisitive'. In moving beyond high tech he now treats each professional task as an opportunity to examine forceful, innovative concepts. He experiments with his design approach on each successive project.

Ushibuka Bridge links three islands in the Amakusa-shoto archipelago in the south of Japan, joining a city centre to its southern suburb. A long-established fishing industry combines here with an exceptionally beautiful natural setting to instil a deeply felt traditionalism in the indigenous population. The designer was commissioned by Hsokawa-san, a former prime-minister and a cultural leader, descended from local nobility. Piano responded to the stringent requirements of sensibility, landscape and technical determinants with the characteristic 'holistic' solution of a trained architect. The pragmatic, engineering approach would be to solve problems sequentially, first tackling the biggest problem and then turning to the secondary and consequential effects in order of priority. Piano sought a single, all-encompassing answer rather than a series of additive solutions. He identified a set of rules (generating principles) that were systematically applied. The

requirements of construction, structural efficiency, road geometry and other factors were given equal weight in the design. The outcome, a box beam of one section throughout, addresses all of these diverse and often conflicting influences simultaneously. There are, inevitably, adverse consequences with this choice but these are either deliberately overlooked or suppressed. For instance, the resulting extra weight is offset by the economy of fabrication. The method operates on the basis that the simpler and more consistent the group of concerns addressed, the greater the clarity of the outcome. The plan and profile of the bridge comprise a huge single bow, curving and rising, connecting three points in a vast geometry. This is not the logic of the highway engineer's transition curves, which are imposed on nature and result in the jarring dissonances of American expressways, but rather a worked line in the landscape that relates to the configuration of space between the islands. The architect refers to the bridge's 'syntactic centrality' within a much larger composition. The bridge could be said

to appropriate its surroundings. The simplicity of the form and structure consciously resists any suggestion that it might be simply a daring mechanical artifact or the expression of mobility. The bridge goes far beyond technical determinism in its efforts to reconcile all the conflicting requirements within a single design move. For instance, the complication of a midpoint access ramp, which needed to be added in the second stage of work, was resolved by arranging for an independent spiral ramp with its own scale and integrity to be set up

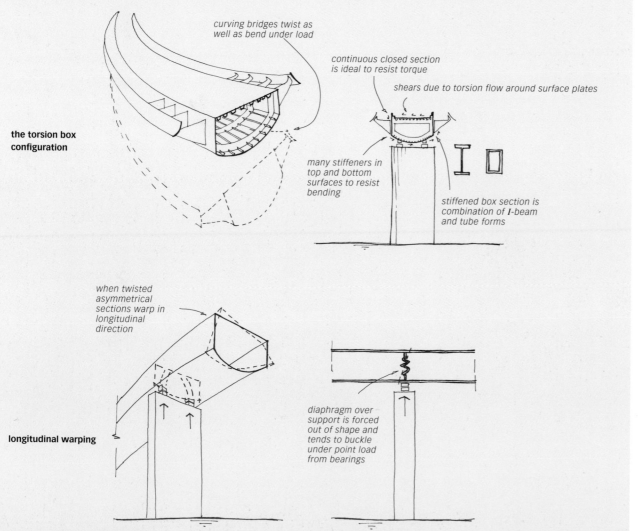

the torsion box configuration

curving bridges twist as well as bend under load

continuous closed section is ideal to resist torque

shears due to torsion flow around surface plates

many stiffeners in top and bottom surfaces to resist bending

stiffened box section is combination of *I*-beam and tube forms

when twisted asymmetrical sections warp in longitudinal direction

longitudinal warping

diaphragm over support is forced out of shape and tends to buckle under point load from bearings

Bridge part elevation

Bridge and site plan

opposite:
The monumentally scaled viaduct crosses the picturesque bay linking the fishing ports of the Amakusa-shoto archipelago.

left:
The torsional resistance and warping characteristics of steel box section girders.

next to the sweep of the main bridge. The central element of roadway is reduced to an extreme simplicity that manipulates the onlooker's sense of scale. The structure is developed as a steel box girder with outriggers supporting footpaths, cycleways and a rim of precast-concrete shield elements. The deep box section is a perfect torsion box that resists the twisting actions of the large outward curves, harnessing the beneficial properties of steel. Its thin plates are butt-welded together and stiffened by internal stringers and cross-frames. A wind tunnel was used to develop the section. The curved soffit

smoothes airflow under the bridge and improves the efficiency of the torsion box. It also reflects the play of light off the waves and generates an optical illusion of lightness; the graduated light pattern appears to lift the horizontal lower edge clear of the supporting piers. This confluence of different actions is typical of a well-handled reductive approach.

The area is prone to high winds, and this needed to be considered in relation to anticipated traffic, which included a large number of pedestrians and cyclists as well as heavy goods vehicles and commuter cars. The concave baffle plates mounted along the upper

edges of the bridge were carefully tuned in the wind tunnel to provide areas of 'stagnation' (almost still air) in their lee, where the footpaths and cycleways are arranged. These repetitive elements catch the shifting light on their upper surfaces and cast changing shadows on the smooth sides of the main beam, emphasizing its continuity. Lightweight components hung onto larger-scale forms recall much earlier Japanese structures that employed masonry bases supporting light timber superstructures and shutters. A minimal number of supporting piers are evenly spaced so that a single structural section

can be efficient throughout the bridge's 900m (2953ft) length. The spans are defined by the size of element that could be prefabricated on shore and then lifted into position by a floating crane. The 150m (492ft) heavily curved units result in a section of unusual depth and physical substance.

The supporting piers are rectangular prisms of reinforced concrete emerging out of the water. They lend the structure gravitas, their arrised forms creating a stark contrast to the curved soffit above. The girder is set off from its supports both physically and visually by extended, spherical bearing pillars. The reticulated shutter pattern on the sides of the piers reflects their form and contrasts with the flush underside of the welded box girder.

Superimposed dead load, extra weight that has to be carried by the bridge structure but which does not help its supporting function, is minimized by the application of some of the designer's earlier innovations in building construction. The wind shields and walkway deck are formed in ferro-cement, an Italian invention and speciality in which precast reinforced-concrete components are pared down by using very light strengthening, typically galvanized chicken wire, dispersed throughout the monolithic section.

The architectural sensibility extends from the sure handling of the bridge's scale right down to the expression of the details. Cast-steel nodes, carefully designed to allow for adjustment, support the baffle plates. Lighting provisions are built in along the plate edges to avoid lamp-posts, which would disrupt the deck line and cause light pollution. Concealed soffit lighting defines the night-time sweep of the bridge across the lights of the bay.

opposite:
Feature lighting emphasizes the overall form and complementary detailing.

left:
The massive lifting capacities of floating cranes were fully exploited in the design and they determined the span sizes.

Petrer Bridge; Petrer, Spain; Carme Piños/Miquell Llorens 1998
12 m / 39 ft

In a run-down suburb of Alicante a dried-up storm gully has been transformed into a landscaped park and play area. A broad deck between the eroded slopes both addresses the topography and plays a central role in the evening promenade, an essential component of Spanish social life. It links two housing estates, providing a new focus and helping to join the two areas into a single neighbourhood. The

bridge is, in effect, a huge element of street furniture. The structure is tiny, no more than a platform across the empty channel that extends on each side into the landscape. Planters for cacti are incorporated into the abutments, and the paving flows without interruption into the surrounding pathways, the timber decking blending into pink sand. The layout gathers up the incoming lines of movement. This

intersection of routes is expressed in an interweaving of three curves: the arch, deck and canopy. The sun provides an additional arc, tied into the structure through the provision of shade. The timber shading is opened in an apparently random pattern that casts an ever-changing chequerboard of shadows across the deck. Formed from a combination of types – half deep-plate girder, half triangulated truss –

the supporting system merges into a curved, ribbed superstructure filled with unevenly spaced planks. The effect is that of a ship's hull eroded by the elements, or the vestiges of an ancient building, perhaps referring to the underlying ruins of an old aqueduct and to the European landscape tradition of Romantic decline. The concave wall draws the gaze towards the distant mountains and captures the light evening breezes, while the open boarding keeps wind pressures mild. The murmur of water is replaced by the sound of wind sloughing through the slats. The movement of the form is increased by the curving plan and sloping ground planes. People walking on the deck are presented first with the close contact of nearby textures and planting and then with distant spaces and atmospheric perspectives. Corners are lifted slightly to separate the route from the seating area. The manipulation of the ground plane is subtle, typical of many modern European architects but also of a fluent landscape architect. Detailing is deliberately over-scaled and laden with references and functions. Huge sections of rolled steel with large web stiffeners emphasize the muscularity of the assembly. The heavy-duty rendering

Bridge elevation

Bridge and site plan

welds contract on cooling and butt welds pull inwards, creating classic 'oil canning' effect

fillet welds draw stiffeners sideways and pull plates downwards resulting in 'hungry horse effect'

the standard butt weld test uses ultrasound which seeks echoes from inclusions, lack of fusion and cracks

some welds are difficult to test because they have too many reflective surfaces

weld distortion and flaw testing

opposite:
The bridge within the landscaped setting of a dried-out riverbed.

left:
Weld distortion and flaw testing.

on the abutments and wing walls is not only physically robust but also amenable to change and amendment. Photoelectric cells, which trickle-charge the lighting, are simply bolted to the superstructure ribs. The detailing reflects the surrounding landscape, which is equally resilient, its spiky cacti and gnarled olive trees epitomizing organic resistance. The assembly does, however, have enough rounded edges to make it suitable for children to play on. The deck is partially supported on ribs cantilevered from the main carriage beam. These elements progressively stick out and curve around, like the remains of a forgotten form. The balustrades are simple, flat handrails and uprights with strainer-wire infills, an absolute reduction of means. Alternate uprights stop short of the handrail as if they have rotted away. The massive deck-level girders have cut faces at each end and appear simply to lie in the planters. The fabricated arches and supporting struts bear on tiny pin connections set among the boulders scattered around each springing. The primitive handling of concrete is exploited in the abutments and retainers; uneven board marks, broken arrises and numerous blow holes give the impression of premature ageing.

The directness of detail and the crude finishing of the installation are intended to sympathize with the industrialized and dilapidated surroundings. Rusting steel, a modified alloy that retains its dark-orange patina, preventing further deterioration, approaches the designer's ideal of a self-finishing product. To give a more subdued effect here, a micaceous iron oxide paint is used to give the steel a dark-grey finish. The paint contains tiny, glistening flecks of mica, which lie in overlapping layers to keep out water and air. It creates a less strident effect than the bright colours of more common protective paints: the zinc oxides, red leads and epoxies. The viscosity of the micaceous iron oxide allows brush marks to stay in the surface, lending the structure a pragmatic, industrial quality. Apprenticed to the late Spanish architect Enric Miralles, Piños extends the practice of taking forms from the immediate landscape, of seeking patterns that resonate with the impulses of nature and drawing texture and feeling from a seemingly barren site.

opposite:
Bridge girders, deck, paving, abutments and planting are all integrated into the spatial design.

left:
The slatted sun and wind shield protects an area defined by slight level changes and boarding patterns.

Swiss Bay Footbridge; Lake Vranov, Czech Republic; Jiri Strasky 1993
252m / 827ft

This is very much an engineer's bridge. An innate aesthetic is achieved through the minimum of means, and the structure satisfies and reflects a specific combination of requirements. It is straightforward and inexpensive but also the outcome of immensely sophisticated analysis and design. Located in a recreation area in the south of the Czech Republic, the bridge links chalets and shops across a lake above a dam, replacing the existing ferry service with a crossing for pedestrians and bicycles. Gas and water supplies run through the pathway. The lake is deep and therefore required a wide single span,

but long, lightly loaded footbridges are an unusual proposition. Footpaths and cycleways are usually clipped onto bigger bridges carrying road traffic, and the self-weight of these structures makes them stable. On wide-span footbridges, however, the imposed load represents the largest part of the overall weight. The distribution of pedestrians as they move across the bridge becomes an important consideration, and the dynamic effects of wind and the rhythmic movements of walking are critical concerns in the selection of economically sized parts.

Eastern Europe has a strong engineering tradition. Sound academic backgrounds and analytical approaches are underpinned with practical and open-minded attitudes to construction and detailing. In this scheme, the bridge engineer Jiri Strasky achieves a simplicity of form that belies the sophistication of its engineering. Sagging cables are sheathed within a precast deck and then prestressed against the deck itself to create a self-stiffening whole. Dispersed loads and dynamic forces are spread through an extremely thin deck assembly. Strasky employed the same integrative approach

and clever manipulation of construction phases used in the development of the stressed ribbon form to devise the extraordinary configuration used at Lake Vranov. Initially, he examined the feasibility of four systems. An extended cantilever proved too heavy, and a tied-arch structure threatened to dominate the surroundings. Similarly, a cable-stayed configuration would have resulted in pylons towering above the surrounding trees. A long, low suspension bridge offered the least intrusive option. The main cable profile is predetermined by the span and the pylon height. Shallower curves lead to greater

flexibility, and therefore compensatory deck stiffening must be added to sustain uneven loads. The standard method on large bridges is to add deep deck trusses, although various trussed hanger arrangements have been experimented with as an integrated alternative for smaller bridges. Another option would be to add a streamlined, steel box girder deck, but it proved too expensive for this scheme, which was to be made entirely from reinforced concrete, the established economic building material of the country. A fourth method of increasing rigidity is occasionally seen on light pipe bridges,

in which opposed cable lines are set up in order to steady the central element within a tensile web. The presence of utilities in this bridge called this method to mind, and the form was adopted. At Lake Vranov the combination of low bank levels and a high centre clearance for sailing boats resulted in a curving deck profile that enables the deck to act as a tie-down. An inverted 'stressed ribbon' is integrated with a traditional suspension bridge arrangement to create a structure whose stiffness comes from interaction between the parts. The very long narrow bridge requires horizontal stiffening to sustain the

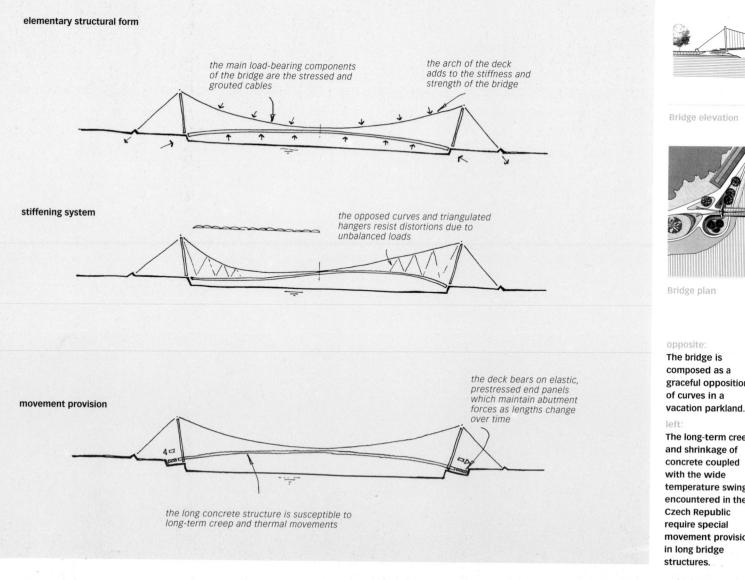

elementary structural form

the main load-bearing components of the bridge are the stressed and grouted cables

the arch of the deck adds to the stiffness and strength of the bridge

stiffening system

the opposed curves and triangulated hangers resist distortions due to unbalanced loads

movement provision

the deck bears on elastic, prestressed end panels which maintain abutment forces as lengths change over time

the long concrete structure is susceptible to long-term creep and thermal movements

Bridge elevation

Bridge plan

161

anticipated wind loads. Three solutions to this problem were examined. Splayed pylons result in inclined cables, introducing sideways components of force that resist wind loads. Alternatively, horizontal cables can be fixed at walkway level, or the deck itself can be widened towards the supports. Strasky, considering Y-shaped pylons to be unsightly and that additional horizontal cables would add clutter and expense, opted instead for a waisted deck beneath A-form pylons. The pylons and hangers were set up perpendicular to the bridge deck axis for visual simplicity. The vertical profile of the deck

and the horizontal plan shape are set out as arcs for ease of construction. To reduce the depth of the deck, pipe diameters were made smaller by dividing the service provisions into two equal halves housed in aerodynamic fairings outside the structural core. Only a small amount of space is available for the backstays at each end of the large central span. A self-anchoring system was reviewed, in which the tensions in the stay ends would feed back into the deck, enabling it to act as a strut. The foundations would then be designed to take only vertical loads. In such a system concrete shrinkage would

seriously compromise prestress levels, and cracks would appear. Instead the bridge employs a partly self-anchoring system, for which end blocks were anchored into the ground and pre-tensioned to the adjacent pylon bases. The pylons were then cast horizontally on the ground before being tilted up. A tie between the saddle points kept the assembly in shape while the main suspension rods were run out, with the hangers already in place. The build-up of these ties had been refined in earlier projects by Strasky, who was especially concerned about long-term durability. The hangers are fixed to a metal sleeve

through which parallel strands are pushed and which is eventually grouted to make it solid. The deck was provided with internal tendons and run out in sections, precast for durability and pinned to each hanger for ease of erection. Once the entire system had been loosely assembled, which left gaps in the main cable sheathing, the deck panels were jacked against one another. The internal tendons and strands were prestressed against the anchor blocks, the main cable sleeves were welded and grouted and the deck segments packed. When the deck-level tendons were finally cut their tensile load was

transferred onto the deck panels, resulting in concrete compression maintained under all loads. The upper strands were simultaneously released to compress their grout cores and sheathing. The concrete remains in compression during shrinkage and thermal movements, the cracks squeezed closed to maximize durability. As the bridge ages and the components change size the proportions of load supported by the cable and the deck will move around. Accepting this process and guiding it within the bounds of what is acceptable during the life of the structure is a sophisticated engineering concept. An extended

collection of static and dynamic tests was undertaken to prove the design assumptions and the feasibility of construction. The very low natural frequencies of the finished structure are below pedestrian excitation levels. High levels of internal damping keep deck-level accelerations down to perceptible but not impassable levels in all but the strongest of winds.

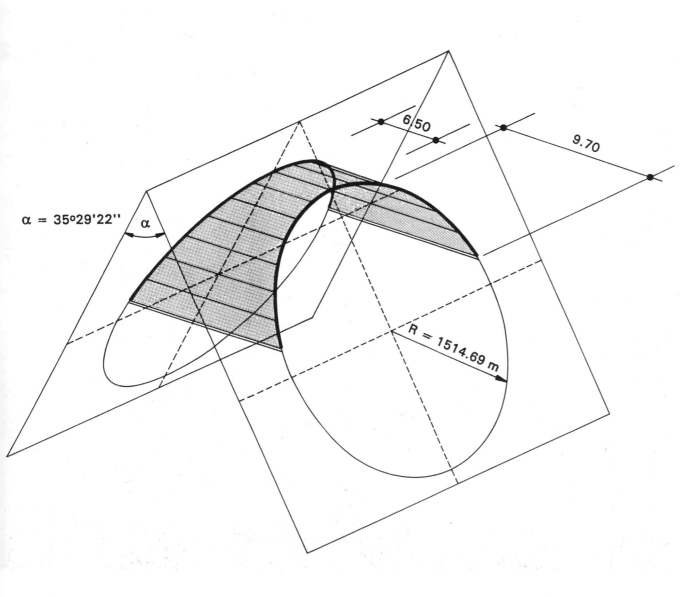

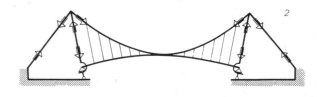

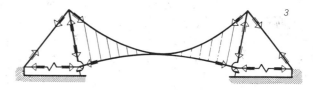

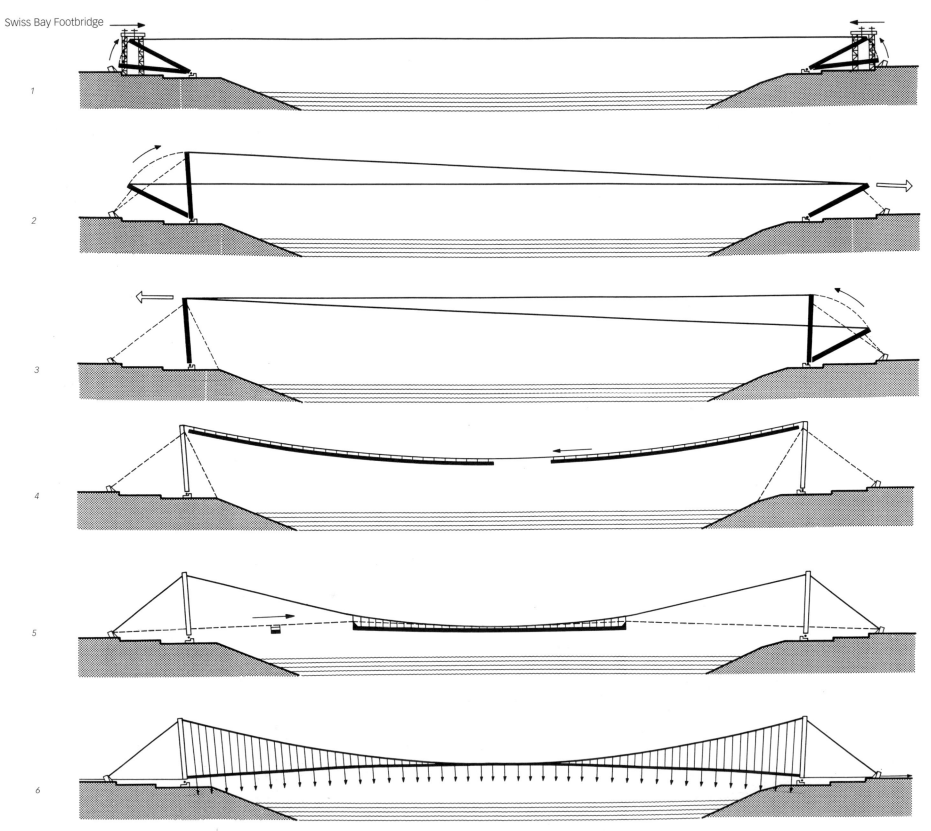

opposite:
Erection sequence:
1. A temporary link was established.
2. The western pylon was tilted up.
3. The eastern pylon was raised.
4. The main cable duct was temporarily suspended in position.
5. Deck panels were fixed from the centre outwards.
6. The system was stressed.

left:
A full-scale load test was carried out using military vehicles.

Lechlade Footbridge; Lechlade, Gloucestershire, UK; Studio E Architects / Alistair Lenczner – Ove Arup & Partners 1993
22m/72ft

New buildings and installations inevitably have a social and environmental impact and are therefore almost always subject to some form of regulation. Planning permission in England seeks to protect the interests of the largest number of people and, as a result, any alteration to a sensitive location can lead to protracted debate. One method of gaining greater acceptance for a proposal is to hold a design competition. With a spread of suitably qualified judges it should be possible to select a design that appropriately reflects the *zeitgeist* and meets the highest standards of current practice. An open design competition was staged in the early 1990s for a footbridge near

Lechlade on the River Thames. The Thames Footpath winds through town and country, a long-distance walkway stretching from the capital city into middle England. This new crossing along the route would open up an attractive stretch of the river to the public and allow walkers to avoid using a hazardous road bridge.

The competition received as many as 600 enquiries and nearly 200 entries, reflecting an overwhelming interest among British designers, and was won by the Polish émigré architect Cezary Bednarski. His design was for a footbridge only 22m (72ft) long, smaller than a standard road crossing. At this scale it was possible to propose a

completely prefabricated structure. Loads up to 25m (82ft) long can usually be carried without escort on British roads, and the anticipated weight of the structure was within the capacity of readily available mobile cranes. By making the most of prefabrication the disruptions of site work would be minimized; the bridge could literally appear overnight. Its construction would also conform to the modernist preference for factory-produced building, and standards of finish would be the best that quality assurance procedures can achieve.

Bednarski's design has a clarity and simplicity suited to its site. It consists of a single arching span with a graceful crescent of spandrel

panels subtly angled to catch the light. The balustrades employ canti-levered glass, with toughened glass panels rising from a clamp plate hidden in the deck edge to an attenuated handrail. The system was invented for the approach ramp to Foster's Sainsbury Centre, Norwich, and creates little visual intrusion. Bednarski's work is characterized by such classic purity, usually combined with particular visual references. In this case he draws on the low, arching rainbow bridges of ancient China and the engineering simplicity of English canal bridges.

The freedom and uncertainty attached to open competitions have led to a tradition of experimentation among entrants, in which the exercise is treated as an opportunity to explore new ideas. The requirements at Lechlade for a small-scale bridge with a simple form, combined with an adequate budget, encouraged the designer to develop a scheme that uses carbon fibre composites as the basic structural material.

Reinforced plastics have occupied a marginal position in building science since their inception in the 1950s. Despite the fact that they offer very high strength to weight ratios and are quite elastic and flexible, they are regarded as extremely expensive by an industry still dominated by traditional methods.

The Lechlade Bridge is a well thought out application of the material. The form is incredibly light, the simple, soft-edged shape reflecting the way it is made. The box girder could be readily formed from reinforced plastic, and the material's strength is so high that wall thicknesses could be minimal. To avoid local buckling the highly stressed skin was to be laid up on an expanded plastic core, a two-part, integrated system pioneered for use in boat hulls, surfboards and glider wings.

Plastics are generally soft and susceptible to degradation by ultraviolet light, and to counteract this composites have traditionally

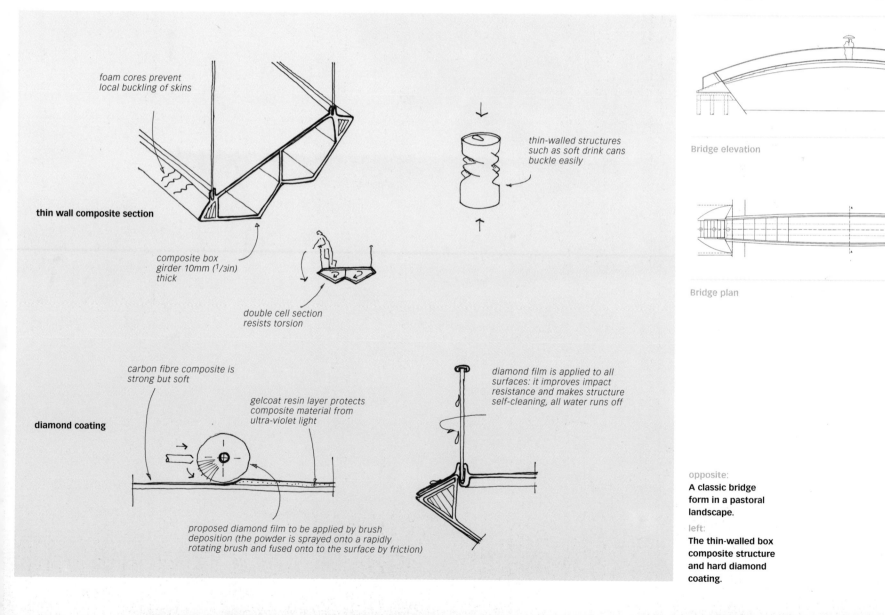

foam cores prevent local buckling of skins

thin wall composite section

composite box girder 10mm (1/3in) thick

double cell section resists torsion

thin-walled structures such as soft drink cans buckle easily

carbon fibre composite is strong but soft

gelcoat resin layer protects composite material from ultra-violet light

diamond coating

proposed diamond film to be applied by brush deposition (the powder is sprayed onto a rapidly rotating brush and fused onto to the surface by friction)

diamond film is applied to all surfaces: it improves impact resistance and makes structure self-cleaning, all water runs off

Bridge elevation

Bridge plan

opposite:
A classic bridge form in a pastoral landscape.

left:
The thin-walled box composite structure and hard diamond coating.

167

been protected by a resin outer coating. The competition brief called for the footbridge to be as close to maintenance-free as possible, and an extraordinary finish was therefore proposed, in the form of a thin diamond coating applied by brush deposition, a technique invented and made feasible by the Hungarian inventor Ernest Nagy. To make the coating, diamond powder is blown onto a rapidly rotating brush that rubs a microscopically thin layer into the surface of the plastic. Since very little material was used, the whole bridge would have required less than 3kg (7lb) of industrial diamond powder. After application the face of the structure becomes not only rock hard but

also, since water slides off the diamond pattern, self-cleaning. A diamond coating to the glass balustrades of the bridge was to have completed the protection system.

A strand of modernist design thinking believes that the future of design will be defined by the introduction of such new materials. In the same way that steel transformed building, leading to high-rise city centres, so plastics will generate new forms. The assumption is that the more high-tech the material, the better its performance. A growing pressure to reduce energy consumption and the argument that design should match performance to requirements, rather than improving

performance to facilitate new requirements, may, however, moderate this enthusiasm.

Although the Royal Fine Arts Commission, a government-appointed panel of experts, praised the Lechlade design, the judging panel raised misgivings about the economic feasibility of the material despite its careful use, and a group of concerned local residents went further, denouncing the scheme entirely. One protester labelled the design 'a yuppie tennis racquet from hell', alluding presumably to the high-tech material. The form of the bridge might have been innocuous but as a whole it did not match a set of local expectations. As a result,

the design was never put into production and the scheme was abandoned.

Other experiments in introducing new plastics into bridge design have been more fortunate. Aberfeldy Footbridge, 1992, which is situated on a Scottish golf course, takes a standard configuration of cable-stayed bridges and applies plastics technology to each part. The main stiffening beam is carbon fibre, and the stays are parafil ropes. These cables of drawn polyaramid, now approved for use as deep-mooring cables in the offshore oil industry, will undoubtedly make an important contribution to bridge development in the future.

Similarly, a footbridge adjacent to the historic cast-iron bridge at Coalbrookdale, in Shropshire, proposes a carbon fibre structure, making direct reference to the pioneering nature of Abraham Darby's original. The designers of this new footbridge endeavour to use the material in new and unusual configurations, emphasizing the fact that the bridge employs a new material in a new way. The structure invites direct comparison with the original bridge, which was heavily constrained by the limitations of its age: the parts are iron and the details all stock forms of carpentry. The Aberfeldy Footbridge tests the new materials within an established format to prove and advertise their worth, whereas the Coalbrookdale proposals set out to explore the potential and consequences of adopting composites in a spirit of idealism. It is extremely unfortunate that the Lechlade bridge, using the most sophisticated technology in an elegant and unmannered way, could not be built in order to exert an even stronger influence on future developments.

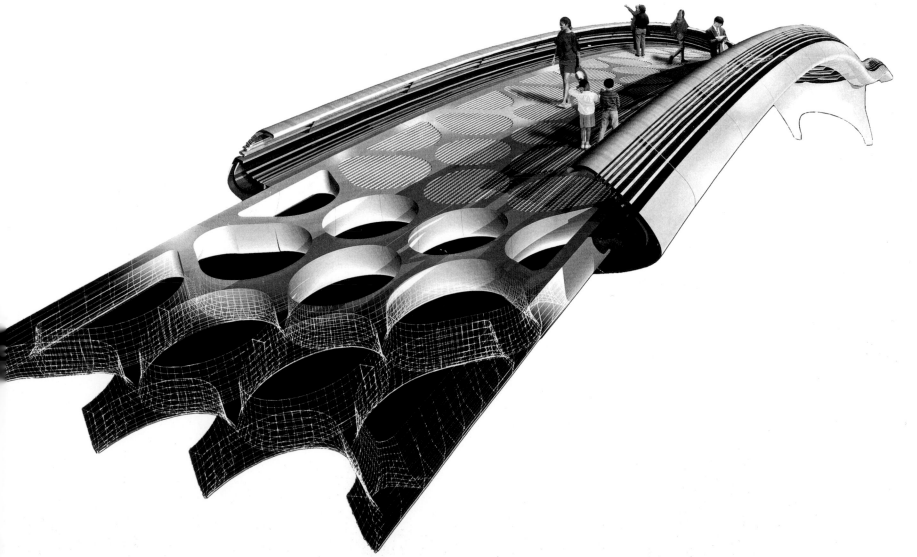

opposite:
For the Aberfeldy Golf Course Footbridge, designed by Maunsell, plastic composites were used in a traditional format.

left:
Coalbrookdale bridge: a proposal exploring the particular qualities of the new materials.

Erasmus Bridge; Rotterdam, The Netherlands; Ben van Berkel/Dept. of Public Works of the City of Rotterdam 1996
800m/2,625ft

Advertisements, postcards and tourist brochures present Erasmus Bridge as a symbol of the city of Rotterdam. The bridge connects the city centre to the dock area of Kop van Zuid, the focus of an intensive regeneration programme, and its unique form contrasts strongly with the sprawling, undifferentiated surroundings.

The design team, including the architect Ben van Berkel and the city's public works department, has a track record of collaboration on buildings that explore expressive form. The hierarchy of a building group headed by the architect inverts the standard practice of bridge-building, in which the engineer leads, and this is reflected in the Erasmus scheme, which is dominated by formal considerations. Significantly, the span is not so great that technical requirements constrain the design.

The structure adapts to its siting off the end of an island with an asymmetrical arrangement. On the west side a classic fan configuration of cables supports the main span from the cranked pylon. To the east backstays are tied down to an intermediate abutment. This pier houses the bearing of a lifting bridge, the largest bascule in Europe, which allows the passage of large ships. The weight of the heavy plate girder lifting section with its massive bearings and counterweight acts as a balance to the main bridge.

The bridge has a studied poise, an anthropomorphic quality that indirectly reflects the forces within. The shaped tower is the result of logically working through a structural premise: bending forces in the asymmetrical pylon needed to be reduced as much as possible, and

the upper cable ends had to be spread out to provide room for their anchor blocks. The cable configuration is laid out to converge on virtual intersection points. Focusing all the forces onto these abstract points in space gives the tower its bent form and directs the whole weight of the structure straight down to the bearings.

Both sides of the A-frame are separated throughout their height so that each supporting fan of cables is aligned into a distinct field of force. These planes intersect at the apex where the form is inflected

to accept the large anchor blocks of the two main backstays. The median split is carried through as a crease across the top plate.

The mast was originally to have been reinforced concrete but was ultimately made from steel to reduce handling weights. Thick plates are hard to press into curves without cracking and so the box components are sharply arrised. The sculpted A-frame loses some of its feel from not being moulded, and its smooth planes appear flat and hard in the even coastal light. The overall height of the structure was successively

reduced to the simple proportion of 1:2, tower to deck, an efficient ratio but one that lends a compactness to the upper works.

Secondary measures have been taken to adjust the visual impact of the pylon proportions. The steel is painted a light blue, allowing it to blend into the North Sea skies like a camouflaged warship, and the ethereal cable fans supporting the main span readily disappear into the haze. The backstays are several degrees more pronounced, and therefore a visual imbalance tends to appear in overcast conditions,

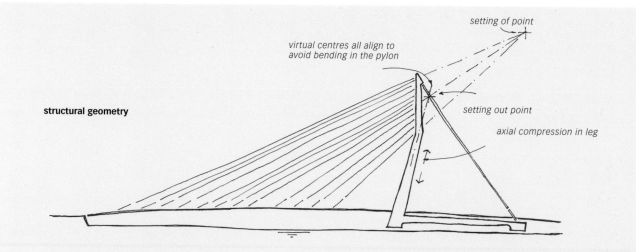

structural geometry

setting of point

virtual centres all align to avoid bending in the pylon

setting out point

axial compression in leg

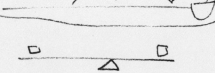

balanced form

mast-head cable anchorages

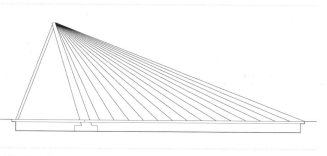

Bridge elevation

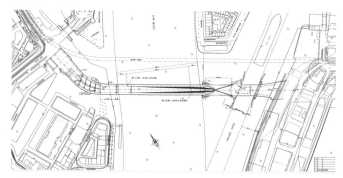

Bridge plan

opposite:
The bridge against the backdrop of the modern city of Rotterdam. The bascule section is raised for a ship's transit.

left:
The pylon's geometry is set out around virtual centres to reduce bending forces and to leave enough room for the main stay anchorages.

171

but night-time feature lighting redresses the problem. The supports are rendered as edges of light from which rays of cables reach forward and down.

The fine parallel-strand cables of the fan stays are susceptible to flutter effects from the winds coming off the North Sea. These movements would induce fatigue in the anchorages of the hardened-wire stays and therefore dampers have been added in the form of rubber socks around the base of the cables, housed in beautiful spun and brushed aluminium cases. Additional damper stays connected to shock absorbers are added underneath to make an expressive deck-level detail.

The details of the bridge as a whole are studied and consistent. The architectural concept of symmetry is carried through the entire design: each part reflects the whole. The lamp standards are miniature versions of the main pylon, and the concrete supporting piers, railings and joints share common forms.

The bridge is founded on driven steel piles in the shallow River Maas. The long steel stems were readily hammered into the lower strata from barges and provide a strong and resilient foundation, capable of sustaining ship strikes. The overall symmetry of form is carried through to the concrete cappings and starlings, but timber fendering, added to ease the passage through the lifting section, forms a discordant note. It is difficult to stop additions detracting from such an integrated design.

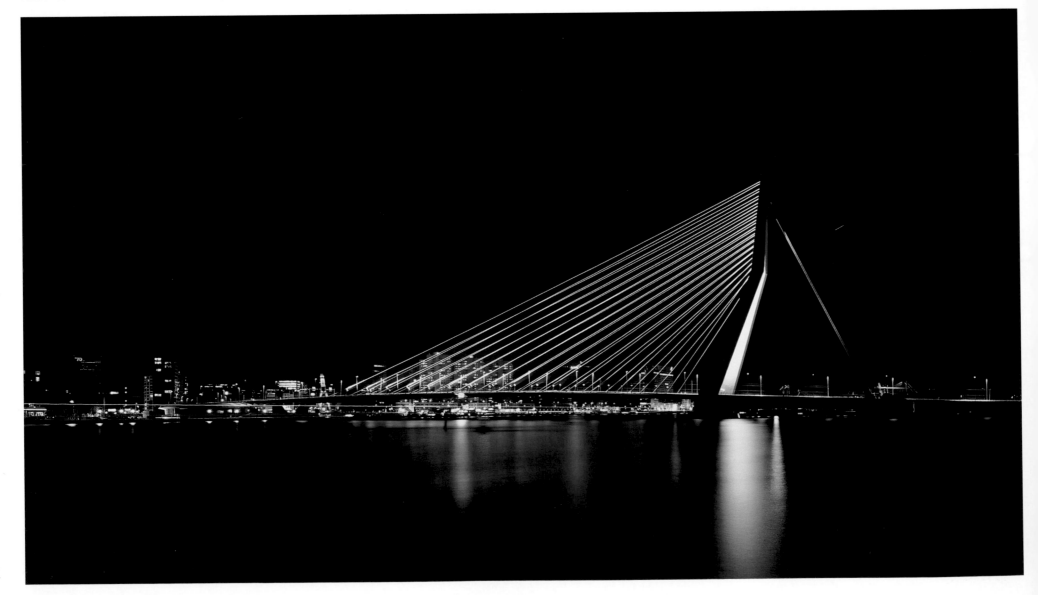

In the construction of the bridge advantage was taken of the location to use massive floating cranes developed for marine salvage and dock work. Prefabricated box elements were assembled horizontally and then rocked up to their vertical positions. By setting up and fabricating at ground level in sheltered conditions, weld quality is improved, and temporary works – access platforms and shores – are minimized. The configuration of the compression struts, backstays and abutments is clearly influenced by the process of erection.

Erasmus Bridge has achieved the landmark status it sought and has become a symbol of the revitalized port. Its modern maritime feel is achieved by subtle means: the development of the structure does not refer directly to the large handling cranes that now epitomize the dock areas, but there is a resonance that comes from the correspondence of underlying form. The ship-building expertise that was on hand influenced procurement, creating an artifact that ties in with surrounding structures. In this way, the Erasmus Bridge manages to combine a particularly idiosyncratic form with an appropriate response to a very distinctive context.

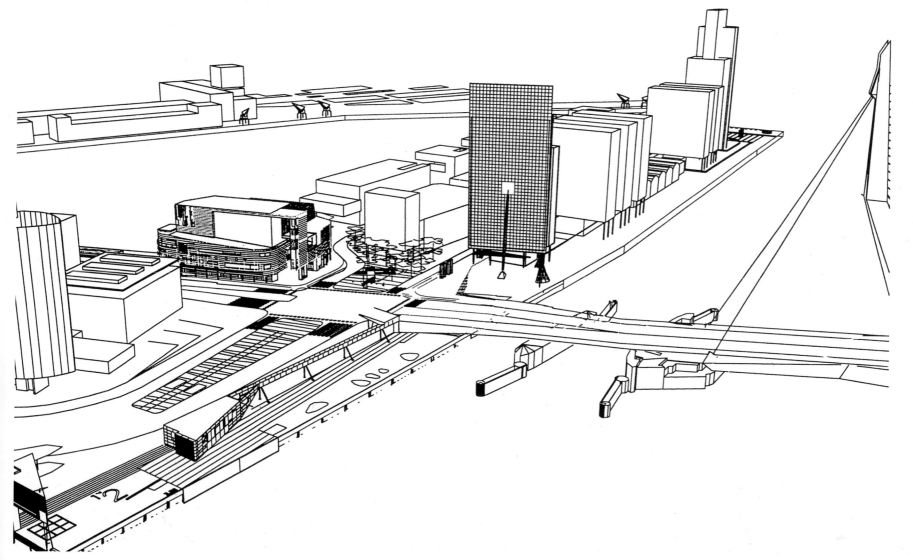

Pont de Normandie; Honfleur, France; Charles Lavigne/ Michel Virlogeux – SETRA 1994
856 m / 2808 ft

The completion of the Pont de Normandie represents a quantum leap in the development of cable-stayed structures, and as a result has brought about a fundamental shift in modern bridge design.

Before the Normandie's construction, suspension bridges had been the proven solution for spans greater than 610 metres (2000 feet). Their long draped cables, stiffened by deck girders or roadway profiles shaped to hold steady in rising winds, have plentiful capacity for crossings in excess of one kilometre (half a mile). The construction method of longer suspension bridges follows a tried and tested sequence founded on the technique of 'air spinning': the necessary giant cables are made *in situ* from a bundle of much finer wires run back and forward across the void.

By contrast, cable-stayed bridges are simpler, contain fewer parts and, if carefully detailed, are easier to maintain. The high degree of redundancy in such structures, that is to say the many different ways that forces can distribute themselves, means that critical components can be replaced without having to shut the bridge down. Compared with suspension bridges, cable-stayed examples are more straightforward to construct. Working outwards from the towers sections are added with their supporting stays until the final connection is made across the middle. As construction proceeds the bridge deck acts as a prop pushing the cantilever out towards the centre. Forces build up with each successive addition, and the growing structure strains and distorts. This is accounted for by pre-setting the various lengths of structural element. Allowances are made so that the parts will be the right length in their final condition once all the stretching has taken place. Minor adjustments are then made in the stays to ensure that the two halves of the bridge meet.

This straightforwardness and adjustability suits bridge-building contractors. Profitability and competitiveness are not only dependent

on efficient working but also on the management of the large-scale risks involved. The limiting factors on cable-stayed designs for the widest crossings are the huge compressions generated in the decks and the control of alignments and forces during construction.

By the mid-1980s conditions were ripe for a leap in the scale of these bridges. Various regional development plans around the world identified sites on river estuaries that required crossings. These projects involved the need for wide spans that could be navigated by large ships and sites that were surrounded by miles of soft alluvium unsuited to support the enormous anchor blocks that suspension bridges require. The crossing of the Seine at Honfleur near Le Havre, Normandy, presented just such a set of conditions.

The French engineer Jean Muller, a protégé of Freyssinet, had built several medium-sized cable-stayed bridges, refining details and improving construction techniques. This series of smaller technical advances paved the way for a cable stay configuration of much greater magnitude. In the last twenty years large-scale structural monitoring systems have become commonplace. For example laser sighting – electronic optical instrumentation designed to tell where an object is in space – had become an industry standard, gaining widespread acceptance for its speed and simplicity of use. A major part of construction revolves around accurately setting out a structure and then checking tolerances as work proceeds. It is natural enough for something to go awry as it is built up. Under load the constituent material strains and therefore the whole structure deflects slightly – a minute and imperceptible amount in a building or a well-designed chair, a good deal more, though still barely perceptible, in the middle of a large-span bridge. These deflections also reveal the underlying stress conditions within the form itself. If one can tell exactly how far a structure has moved from its original resting point then the internal stresses can accurately be

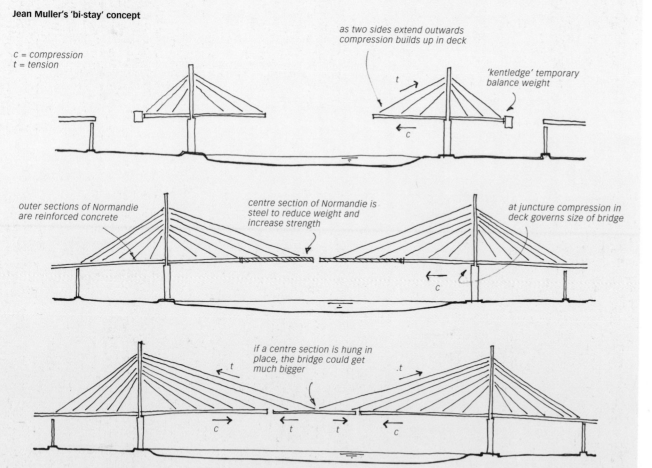

Jean Muller's 'bi-stay' concept

c = compression
t = tension

as two sides extend outwards compression builds up in deck

'kentledge' temporary balance weight

outer sections of Normandie are reinforced concrete

centre section of Normandie is steel to reduce weight and increase strength

at juncture compression in deck governs size of bridge

if a centre section is hung in place, the bridge could get much bigger

Bridge elevation

opposite:
The wide span of the cable-stayed bridge above the extended alluvial estuary of the River Seine. The backstays are secured down to closely spaced piers beneath the approach ramps.

left:
The influential engineer Jean Muller's bi-stay concept for extending the practical span of cable-stayed bridges.

175

deduced. The safe construction of a large bridge, if it is not to be over-strained, becomes a matter of building it within an envelope of space beyond which it must not stray. During construction the Pont de Normandie was monitored, reanalyzed and adjusted every day so that a pair of stays could be added to each end every twenty-four hours.

Speed is paramount for economic bridge construction. A great proportion of the overall cost, sometimes as much as fifty per cent or more, is not in the built fabric at all but in the so-called preliminaries: money for the establishment and maintenance of the small city of workers needed for the building operation; for the hugely expensive, sometimes specially designed and constructed plant; cranes, platforms, jacks and monitors; for insurance and of course for the cost of the capital tied up in the enterprise before the structure can be opened and a return made. The construction speeds achieved on the bridge at Honfleur made it enormously influential on the choices of system on subsequent projects.

The build-up of forces in the giant cable-stayed structure was controlled by the use of a mixture of materials. The approach causeways were economically formed of prestressed concrete box beams, a technology invented in France and subsequently improved upon by French engineers. These robust structures formed firm anchorage points for a central deck girder of high-strength steel – relatively light, strong and stiff but with the potential for extensive, safe adjustment. The regimented nature of the French engineering bureaucracy is such that a few years previously such a hybrid construction of disparate materials would have been extremely unlikely. The designer of the bridge, Michel Virlogeux, had amalgamated two independent government departments – one responsible for concrete bridges, the other for steel – in the face of entrenched resistance. The Pont de Normandie is an almost deliberate display of the advantages of not differentiating between the two construction forms.

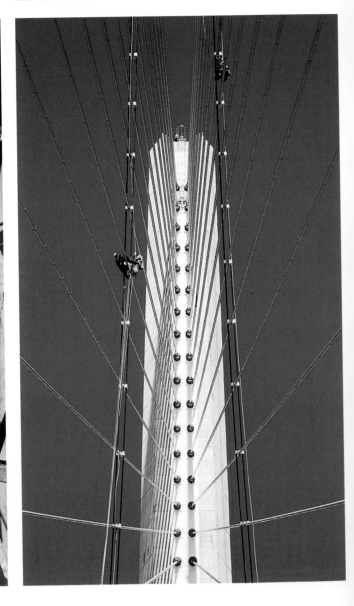

At the time of its completion the Pont de Normandie was the longest cable-stayed bridge in the world and so exceeded the previous record it looked set to hold its place for some time. It is, however, now surpassed by one of the bridges in the Honshu-Shikoku project in Japan, which is some 30 metres (98 feet), or one per cent, longer.

Michel Virlogeux has outlined his approach to the aesthetic component of bridge design in several interviews, but has been roundly criticized by other engineers for raising such issues at all. For example he recalls how when his first project was completed he realized it looked 'a horror'. Aesthetics were important in Virlogeux's work, but where they were meant to come from was unclear. Virlogeux's first response was, therefore, to attenuate everything, to remove millimetres wherever it was safe to do so. Many engineers will recognize this reaction. Then he began to call upon architects to collaborate, wisely deciding that the design of form is something that requires a dedicated architectural education and taking pride in the different consciousnesses of architect and engineer. He laments, however, that the best French architects seemed unable to collaborate with engineers at all.

Virlogeux's masterpiece – the culmination of his development over several projects – has proved photogenic and influential. The simplicity of the Pont de Normandie demonstrates its engineering rationale. The bridge is big and long because the traffic demands on it are light. The ranks of approach piers reflect the engineer's cautious reaction to the scale of the undertaking. The A-frame pylons and extended tops seem slightly mannered, as if the engineer was still uncertain about what an appropriate aesthetic should be, but the slender proportions and the ethereal ties, which seem to blend into the Normandy light, are perfect.

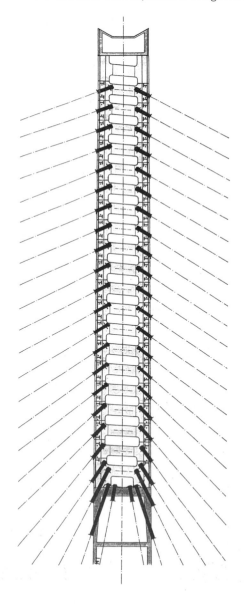

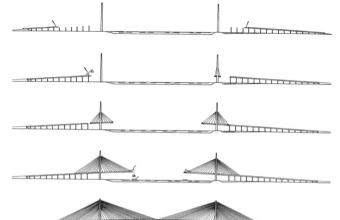

opposite, left:
The climbing formwork platform and access scaffolds on the pylon anchor blocks during construction.

opposite, middle:
Each main stay comprises parallel strands of wire drawn individually into the bundle up to the upper anchorage.

opposite, right:
The cables are prevented from vibrating in the wind by light cross-ties.

far left:
The numerous stay anchors are set out in a well-designed anchor box reinforced with heavy cross straps.

left:
The bridge construction sequence showing the massive tie-down provisions in the backspans.

below:
The main stays are protected from local bending by pin ends and sock dampers just above the anchor blocks.

opposite:
Jean Muller's Brotonne bridge over the Seine (1974). Precedents such as this bridge allowed French engineers to develop the cable-stayed concept on an unprecedented scale, so that it was competitive with all but the very largest suspension bridges.

left:
Roadway lighting suspended from the main cables; a rare configuration previously avoided due to uncertainties about cable behaviour.

Gateshead Millennium Bridge; Gateshead, UK; WilkinsonEyre Architects / Gifford and Partners 2000
105m / 345 ft

Many of Britain's northern cities are established along rivers that once formed a central resource for trade and development. The decline of heavy industries left blighted waterfronts that are now the target of urban regeneration programmes. Central and local government initiatives have appropriated the symbolic potential of bridges to make these structures the centrepiece of many projects. As part of the redevelopment of East Gateshead, a depressed suburb of

Newcastle-upon-Tyne, an arts complex and a regional music centre now replace the ship-building area, while across the water riverside cafes, restaurants, hotels and shops thrive on the edge of the city centre. The Gateshead Millennium Bridge contributes an element of drama to this setting. The competition to design a footpath and cycleway linking the north and south banks of the River Tyne was won by the architects WilkinsonEyre in 1995, with a scheme that responds

directly to its surroundings. The arch of the bridge refers back to the famous Tyne Bridge, only 300m (984ft) upstream. The precursor of the Sydney Harbour Bridge, the old Tyne Bridge stands within a group of Victorian swing bridges and pier heads that have come to epitomize Newcastle. Its high tied arch and monumental abutments convey a feeling of repose. The Millennium Bridge takes the structurally efficient profile of the old arch bridge and tilts it over. The lateral forces induced

by the tilt are counteracted by curving the suspended deck on plan. The bridge extends the riverside walk downstream from the busy pier head. Pedestrians are taken out to a view of the Baltic Mill (which houses the arts complex) from the middle of the span and then back towards the new centre of activity on the Gateshead side. The old bridge upstream is viewed from between the suspension cables and beneath the overhead arch.

A significant air draft is required to allow ships to pass along the river, especially when the Tall Ships Race convenes periodically on the Tyne. The bridge follows the 'sickle bridge' format explored by the Spanish architect Santiago Calatrava, but is also openable. Calatrava's 'eyebrow' metaphor, used in one of his kinetic sculptures, is appropriated and transformed. The walkway is given a pronounced bow and the whole assembly is set on two hinges. The supporting arch

can be lowered and the deck raised until their crowns reach a common level. The intervening cable ties become horizontal, clearing an adequate passage for masted vessels.

The movement is subtle, the centre of gravity of the whole system remaining constant as the descending arch counterbalances the rising deck. No mechanical effort is expended on lifting the assembly and so power is only required to keep the structure stable. The curving

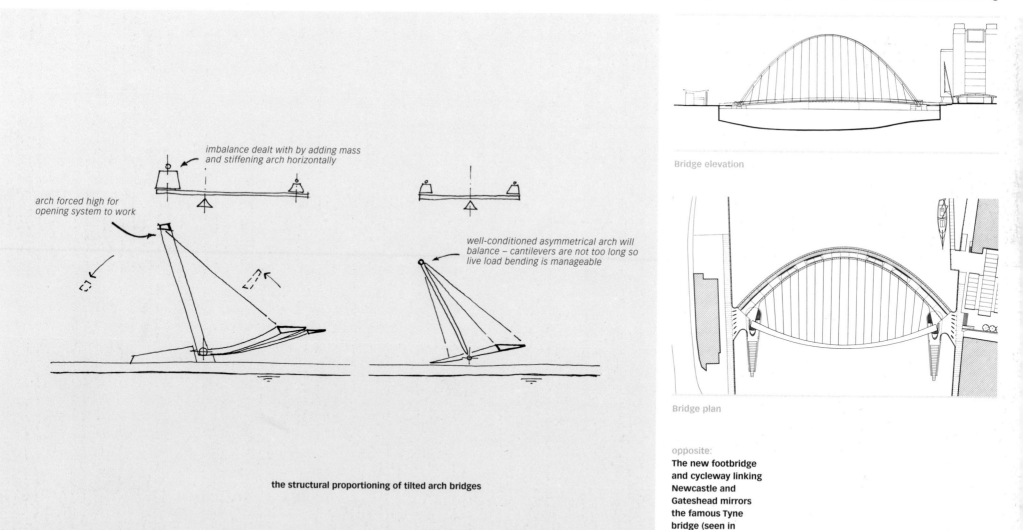

imbalance dealt with by adding mass and stiffening arch horizontally

arch forced high for opening system to work

well-conditioned asymmetrical arch will balance – cantilevers are not too long so live load bending is manageable

the structural proportioning of tilted arch bridges

Bridge elevation

Bridge plan

opposite:
The new footbridge and cycleway linking Newcastle and Gateshead mirrors the famous Tyne bridge (seen in the distance).

left:
The balance of tilted arch bridges.

movement is graceful and the transformation dramatic. Long hydraulic rams extending upstream from each pier form the actuators. The machinery is set in glass engine houses, a curious revival of the Victorian taste for displaying large mechanisms.

The bridge's dramatic effect is achieved at some cost, however. The curves of the deck and the arch are forced out as far as is practical but they are not in equilibrium and must be balanced by a concrete filling concealed within the arch crown. Wind loads inevitably bend the free-standing arch but the applied loads of pedestrians and cyclists also pull it out of shape. The grace of a true parabolic arch, its forces pulled into a funicular line of compression running within the member, is therefore lost and a stocky box-section of steel takes its place.

As a result, the structure weighs 840 tonnes (827 tons), 8 tonnes per metre of span (over 2 tons per foot) and five times as much as a conventional structure. This factor may account for the visual disparity between the competition rendering and the bridge itself. The phenomenon of 'corporality' is another consideration: a built object, perceived by binocular vision, always has more substance and presence than its equivalent on paper.

The selective use of very high-strength materials, hidden components of widely different mass or density, allows for unusual

manipulations of form. The bridge could be seen to be subverting the mind's innate sense of structural logic to induce a sense of excitement.

The structural detailing is concise, the cable anchorages are tidy and the bearings clearly express their purpose and the scale of forces involved. The deck is separated into the footpath and the cycleway by a change in level, to maintain views across the water. The arch and deck cross-sections have been developed to respond to force patterns and to provide the splayed soffits and lighting pelmets needed to light the sweeping curves effectively. Carefully detailed barriers reflect Britain's extensive advances in transport architecture.

The theatricality of the overall concept is crowned by the way in which the structure was erected. The ship-building legacy of the Tyne means that the site can be reached by the largest floating cranes, developed for lifting engines into launched hulls and for massive salvage operations. The whole bridge could be floated upstream from a shipyard assembly point and lifted into place in one piece. Its sudden appearance kick-started the urban regeneration programme, providing an irresistible attraction to the media and promoting interest in the initiative as a whole.

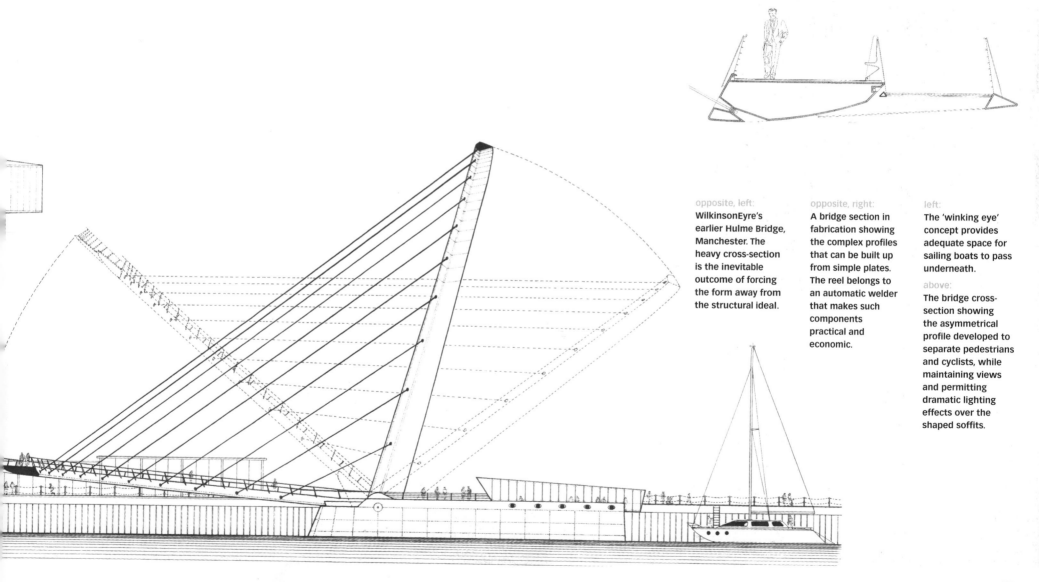

far left:
The erection process developed as a dramatic event to advertise the new crossing.

middle:
As well as lifting the entire bridge, the floating crane was able to transport it a considerable distance from yard to site.

left:
The bridge assembly, supported on a lifting beam, is carefully moved onto locating pins at the site.

Project Credits, Index, Picture Credits, Acknowledgements

Project Credits

Helgeland Bridge
Structural engineer Leonhardt, Andrä and Partner
(Holger S. Svensson) in collaboration with AAS-Jakobsen AS
Client Norwegian Public Roads Administration
Contractor Veidekke ASA in collaboration with Gleitbaugesellshaft
Stay cables Stahlton AG
Wind tunnel tests University of Western Ontario

Plashet Grove Footbridge
Architect Birds Portchmouth Russum
Joint client Education Department, London Borough of Newham
and Plashet Grove School
Structural engineer Techniker
Cost consultant Gardiner & Theobald
Main contractor Spencers
Steel fabricator MSI
Fabric subcontractor Architen

Greater New Orleans Bridge
Main bridge superstructure design Modjeski and Masters, Inc.
Main bridge substructure design F. Lionel Pavlo Engineering Co.
Client Louisiana Department of Transportation and Development,
Crescent City Connection Division
Superstructure contractor Harris Structural Steel Co., Inc.
Substructure contractor Massman-Johnson-James: A Joint Venture

Volantin Footbridge
Credits not available

Wettstein Bridge
Credits not available

Traversina Footbridge
Structural engineer Conzett, Bronzini, Gartmann AG
Client Verein KulturRaum Viamala
Main contractor Flutsch Holzbau
Steel cable Riss AG
Steel joints (Stahlknoten) Romei AG
Concrete contractor V. Luzi

Pùnt da Suransuns
Structural engineer Conzett, Bronzini, Gartmann AG
Client Verein Kultur Raum Viamala
Geology survey Baugeologie
Survey Hasler & Muggler
Concrete contractor V. Luzi
Soil anchors (Ankerarbeiten) Otto Bohr
Steel manufacturer Romei AG
Paving Granitwerk Andeer; A Conrad AG

The Great Belt Link
Architect Dissing & Weitling
Structural engineer COWI Consulting Engineers and Planners
Client AS Storebaeltsdforbindelsen
Consultants B. Hoejlund Rasmussen A/S; Ramboell & Hannemann A/S
Landscape architect Joergen Vesterholts Tegnestue

Millennium Bridge
Architects Foster and Partners, Sir Anthony Caro and
Ove Arup & Partners
Structural engineer Ove Arup & Partners
Mechanical engineer Ove Arup & Partners
Lighting Claude Engle and Ove Arup & Partners
Sculptor Sir Anthony Caro

West India Quay Footbridge
Architect Future Systems
Structural engineer Anthony Hunt Associates Ltd
Client London Docklands Development Corporation
Contractor Little Hampton Welding Ltd
Quantity surveyor Bucknall Austin
Lighting Lighting Design Partnership

Kiel-Horn Folding Bridge
Architect von Gerkan, Marg and Partner
Structural engineer Schlaich Bergermann and Partner
Client Magistrat der Landeshauptstadt Kiel
Moving parts NSO Neptun Stahlobjektbau
Pier construction Heinrich Hirdes

Fred Hartmann Bridge
Architect URS Southern Corporation
Structural engineer Leonhardt, Andrä and Partner
(Holger S. Svensson)
Client Texas State Department of Highways & Transportation
Wind dynamic consultant Dr Robert H. Scanlan
Construction engineer Williams Brothers Construction Co., Inc. and
Taylor Bros., Inc. in collaboration with DRC Consultants
Supply of stay cables VSL Corporation

Steg über der Mur
Architect Günther Domenig; Arch. Dipl.-Ing. Hermann Eisenkock
Structural engineer Prof. Dr. Harald Egger
Client and main contractor City of Graz

Allmandring Footbridge
Architect Kaag & Schwarz
Structural engineer Gustl Lachenmann
Client Universitätsbauamt Stuttgart and Hohenheim

Quantity surveyor Dipl. Ing. W. Zellner
Main consultants Smoltzcyk & Partner; Fa. H. Rothfuss;
Simeth GmbH; Pfeifer

Royal Victoria Dock Bridge
Architect Lifshutz Davidson
Structural engineer Techniker
Client English Partnerships
Main contractor Kier London Ltd
Mechanical and electrical engineer Allott & Lomaz
Lighting consultant Equation Lighting Design
Quantity surveyor Davis Langdon & Everest
Metalwork Darke Tech
Aluminium Alusuisse UK

Talmadge Memorial Bridge
Structural engineer T. Y. Lin International
Client Georgia Department of Transportation
Main contractor Pensacola Tiggert

Charles River Mainline Bridge
Concept Prof. Christian Menn
Management consultant Bechtel Parsons Brinckerhoff
Preliminary engineering (up to 40% stage) Parsons Brinckerhoff
Construction management Bechtel Parsons Brinckerhoff
Final design HNTB Corporation
Client Massachusetts Turnpike Authority
Contractor Atkinson/Kiewit
Steel fabricator Grand Junction Steel
Post-tensioning and stay cable Freyssinet
Construction engineers T. Y. Lin International

Sunniberg Bridge
Client and general management Baudepartement (construction
department) Graubünden
Site supervision Tiefbauamt Graubünden
Project management E. Toscano AG, Chur
Adviser A. Deplazes, Chur
Concept Prof. Christian Menn, Chur
Consultant, technical guidance Bänziger Köppel Brändli and
Partner, Chur
Local technical guidance Wüst Stucki and Partner, Klosters
Proof consultant P. Marti, ETH Zurich
Geological adviser T. Lardelli, Chur
Costruction performance/joint venture Arge Sunnibergbrücke:
Vertsch, Klosters; Preswerk & Cie AG Brückenbau, Siebnen
Subcontracters Form traveller H. Schürer, Zurich
Stay cables/prestressing Stahlton AG, Zurich
Pile foundation Eggstein AG, Luzern

Project Credits

Steel construction AMSAG Stahlbau, Serneus
Drainage system Rowatec, Volketswil
Reinforcement Gisler and Partner AG, Flüelen
Concrete Kieswerk Arieschbach AG, Fideris
Concrete supplier Bündner Cement AG, Untervaz

Solferino Footbridge
Engineer and architect Marc Mimram
Client Ministry of Culture (Ministère de la Culture); Ministry of Public Works, Housing and Transport (Ministère de l'Equipement, du Logement et des Transports) in collaboration with the Louvre (Etablissement Public du Grand Louvre) and the Public Foundation for Cultural Works (Nouvellement Etablissement Public de Maîtrise d'Ouvrage des Travaux Culturels)
Consultant Sogelerg

Sunshine Skyway
Concept Jean Muller
Designers Figg and Muller Engineers (high level and main span); Parsons Brinckerhoff; Quade & Douglas Inc. (low level approach)
Construction management SKYCEI (collaboration of Parsons Brinckerhoff Construction Services; H. W. Lochner, Inc.; DRC Consultants, Inc.; DRC Consultants, Inc. and Kisinger Campo and Associates Corp.)
Client Florida Department of Transportation
Main contractors Paschen Contractors Inc.; Ballenger Corporation

Main Danube Channel Footbridge
Concept, planning and construction Dipl. Ing. Richard J. Dietrich
Structural engineer Ingenieurburo Dr. Ing. Heinz Brüninghoff and Dipl. Ing. Rampf
Client Rhein-Main-Donau AG
Project management Prof Dr. Ing. F. Grundmann
Main contractors Bilfinger & Berger Bau AG; Huber & Sohn Holzbau GmbH
Wind testing Prof. Dr. Ing. V. Denk
Wood testing Technical University of Munich

Roosevelt Lake Bridge
Engineer and architect HNTB Corporation Kansas City, Missouri
Owner Arizona Department of Transportation
Contractor Edward Kraemer & Sons, Inc., Plain, Wisconsin
Erector John F. Beasley Construction Co., Dallas, Texas
Steel fabricator PDM Bridge Corp., Eau Claire, Wisconsin

Miho Museum Bridge
Architect I.M.Pei Architect
Associate architect Kibowkan International Inc.
Engineers Leslie E. Robertson Associates
Client Shinji Shumeikai, Shumei Cultural Foundation

Contractor Shimizu Corporation
Landscape architects Kohseki and Akenuki Zoen; Noda Kensetsu

Ushibuka Bridge
Architect Renzo Piano Building Workshop
Client Kumamoto Prefecture, in association with Maeda Engineering Co.
Consultants Ove Arup & Partners

Petrer Bridge
Architect Carme Piños Desplat
Structural engineer Miquell Llorens
Client and main contractor Petrer Council
Construction NECSO – Entrecanales Cubiertas

Swiss Bay Footbridge
Design Prof. Dr. Jiri Strasky – Strasky Husty and Partners, Consulting Engineers, Brno, Czech Republic and Mill Valley, California
Construction management Ilja Husty – Strasky Husty and Partners, Consulting Engineers, Brno
Construction documentation Jaroslav Jordan and Miroslav Spudil – Dopravni stavby & Mosty, Olomouc
Checking Dr. Marie Studnickova – Klokner Institute CVUT, Prague
Wind tunnel test Prof. Dr. Miroslav Pirner, Prague
Contractor Dopravni stavby & Mosty, Olomouc

Lechlade Footbridge
Architect Cezary M. Bednarski – Studio E Architects
Structural engineer Alistair Lenczner – Ove Arup and Partners
Composite plastics engineer Arthur Webb
Diamond coating Ernest Nagy
Support John Prewer

Erasmus Bridge
Architect Ben van Berkel
Client and technical consultants Department of Public Works of the City of Rotterdam
Lighting design Lighting Design Partnership
Main contractors Heerema Dock Installations, Grootint Dordrecht b.v; Compagnie d'Enterprises CFE S.A., Brussels; N.V. Maatschappij voor Bouw- en Grondwerken, Antwerp; Ravestein-Noell, Deest; Lighting Design Partnership, Edinburgh

Pont de Normandie
Architect Charles Lavigne
Designer Michel Virlogeux – SETRA
Client Chambre de Commerce et de L'Industrie du Havre
Project management Direction Départementale de l'Equipement de Seine-Maritime (SETRA, QUADRIC, SEEE, SOFRESID, SOGELERG, CSTB, ONERA)

Contractors (concrete) GIE Pont de Normandie (Bouygues, Campenon Bernard, Dumez, GTM, Quillery, Sogea, Spie)
Subcontractors Freyssinet, Bilfinger & Berger, DSD/Secometal
Contractors (steel) Monberg & Thorsen
Subcontractors Munch, Sartec, VSL, SDEM

Gateshead Millennium Bridge
Architect WilkinsonEyre Architects
Structural engineer, services engineer Gifford and Partners
Client Gateshead Metropolitan Borough Council
Main contractor Harbour & General/Volker Stevin
Mechanical, electrical and hydraulic subcontractor Kvaerner Markham
Lighting consultant Johnathan Speirs & Associates
Steelwork subcontractor Watson Steel Ltd

Index

Page numbers in *italics* refer to illustrations

Index

Picture Credits

Albert Berenguier (174, 176, 178–9); Courtesy of Water Resources Center Archives, University of California, Berkeley (33); Birmingham Museums and Art Gallery (8 top left); Collections/John D. Beldom (21); Courtesy BPR (52–3); Collections/Gena Davies (41), Collections/David Davis (24 left); Collections/Brian Shuel (40); Peter Cook/VIEW (11 right); Richard Davies (92–3); Richard Dietrich (18 left, 20, 24 right, 25, 140, 142–3); Courtesy Günther Domenig (104, 106–7); Edifice/Darley (15); Edifice/Eddie Ryle-Hodges (123 left); Edifice/Schneebels (14 right); Mary Evans Picture Library (13 left, 23, 35 right, 37); Chris Gascoigne/VIEW (112, 114, 116–7); Richard Glover/ARCAID (90, 94); Courtesy Nicholas Grimshaw (10 bottom); Klaus Frahm/artur (98–9); Courtesy Zaha Hadid (10 top); Robert Harding Picture Library/Adam Woolfitt (18 right); Guy Hearn (Hulton-Deutsch Collection/Corbis (22); Ilja Husty (162); Institute of Civil Engineering (38); Scott Jolliff (118, 120–1); Courtesy Kaag and Schwarz (108, 110–1); Courtesy Leonhardt, Andrä and Partner (44, 46–7); John Edward Linden (61); Duccio Malagamba (156, 158–9); Courtesy Massachusetts Turnpike Authority – Central Artery/Tunnel Project (122, 124–5); Courtesy Christian Menn (126, 128); Hans-Peter Merten/Robert Harding Picture Library (129); Courtesy HNTB Corporation (144, 146–7); Timothy Hursley (150–1); Courtesy Maunsells (168); Courtesy Modjeski & Masters (54, 56); Adam Moerk (Dissing & Weitling (80, 82, 84–5); Moises Puente, Schenk & Campbell (74, 76–9); J. M. Monthiers (130, 132); Courtesy Parsons Brinckerhoff Inc. (9 bottom, 136, 138–9); Powerstock Photo Library/David Taylor (30 left); Powerstock Photo Library (8 right, 31, 32, 36); Noriaki Okabe (155); Christian Richters (170, 172); Neil Rall, Courtesy Mott MacDonald (10 left); Courtesy B. B. Rath, Naval Research Laboratory (29); Franck Robert (48, 50); Paolo Rosselli (58, 60, 63); Schinkenchiku-sha (148, 152, 154); Jörg Schlaich (12 right); Trustees of Sir John Soane's Museum (8 bottom); Harry Sowden (11 bottom); Spectrum Colour Library (40 left); Tiefbauamt Graubünden, Switzerland (26); Jeannette Tschudy (68, 70–3); Victoria Art Gallery, Bath and North East Somerset Council/Bridgeman Art Library (13 right); Courtesy URS Southern Corporation (100, 104); Morley von Sternberg (11 top); Morley von Sternberg/ARCAID (86, 88 bottom), Courtesy WilkinsonEyre (180, 182 left, 184–6); Jiri Strasky (160, 165); Nick Wood/Virtual Artworks (88 top)

Acknowledgements

Author's acknowledgements: The author would like to thank research co-ordinator Jennifer Hudson and project editor Liz Faber at Laurence King, text editor Anthea Snow, and designer Frank Philippin.

Publisher's acknowledgements: The publisher would also like to thank Frank Philippin and Anthea Snow for their hard work and unfailing dedication.

Bridge Span Comparison

Bridge	Span	
Helgeland Bridge	452 m / 1483 ft	
Plashet Grove School Bridge	60 m / 197 ft	
Greater New Orleans Bridge	480 m / 1575 ft	
Campo Volantin Footbridge	71 m / 233 ft	
Wettstein Bridge	66 m / 217 ft	
Traversina Footbridge	48 m / 157 ft	
Pùnt da Suransuns	40 m / 131 ft	
The Great Belt Link (East Bridge)	1624 m / 5328 ft	
Millennium Bridge	162 m / 532 ft	
West India Quay Footbridge	15 m / 49 ft	
Kiel-Hörn Folding Bridge	26.65 m / 87 ft	
Fred Hartmann Bridge	361 m / 1184 ft	
Steg über der Mur	55.8 m / 183 ft	
Allmandring Footbridge	34 m / 96 ft	
Royal Victoria Dock Bridge	127.5 m / 418 ft	
Talmadge Memorial Bridge	337 m / 1106 ft	
Charles River Mainline Bridge	227 m / 745 ft	
Sunniberg Bridge	140 m / 459 ft	
Solferino Footbridge	102 m / 335 ft	
Sunshine Skyway Bridge	365 m / 1198 ft	
Main-Danube Channel Footbridge	73.28 m / 240 ft	
Roosevelt Lake Bridge	329 m / 1079 ft	
Miho Museum Footbridge	120 m / 394 ft	
Ushibuka Bridge	150 m / 492 ft	
Petrer Bridge	12 m / 39 ft	
Swiss Bay Footbridge	252 m / 827 ft	
Lechlade Footbridge	22 m / 72 ft	
Erasmus Bridge	800 m / 2625 ft	
Pont de Normandie	856 m / 2808 ft	
Gateshead Millennium Bridge	105 m / 345 ft	